The Long-Standing Problem of Parasitic Diseases in Zoo Animals: Current Challenges and Searching for Solutions

The Long-Standing Problem of Parasitic Diseases in Zoo Animals: Current Challenges and Searching for Solutions

Guest Editor

Mariana Panayotova-Pencheva

MDPI

Basel • Beijing • Wuhan • Barcelona • Belgrade • Novi Sad • Cluj • Manchester

Guest Editor
Mariana Panayotova-Pencheva
Institute of Experimental
Morphology, Pathology and
Anthropology with Museum
Bulgarian Academy of Sciences
Sofia
Bulgaria

Editorial Office
MDPI AG
Grosspeteranlage 5
4052 Basel, Switzerland

This is a reprint of the Special Issue, published open access by the journal *Journal of Zoological and Botanical Gardens* (ISSN 2673-5636), freely accessible at: https://www.mdpi.com/journal/jzbg/special_issues/3965304B3C.

For citation purposes, cite each article independently as indicated on the article page online and as indicated below:

Lastname, A.A.; Lastname, B.B. Article Title. *Journal Name* **Year**, *Volume Number*, Page Range.

ISBN 978-3-7258-4795-2 (Hbk)
ISBN 978-3-7258-4796-9 (PDF)
https://doi.org/10.3390/books978-3-7258-4796-9

Cover image courtesy of Mariana Panayotova-Pencheva

Contents

About the Editor

Mariana Panayotova-Pencheva

Mariana Panayotova-Pencheva is a highly qualified researcher and academic in the field of veterinary parasitology with over 25 years of professional experience in scientific research, education and project implementation. She is currently an Associate Professor at the Institute of Experimental Morphology, Pathology and Anthropology with a Museum of the Bulgarian Academy of Sciences, where she conducts research in the field of veterinary parasitology. Dr Panayotova-Pencheva obtained her Master's degree in Veterinary Medicine in 1995 and defended her PhD thesis in 2009, in which she dealt with the etiological, epidemiological and pathomorphological aspects of protostrongylidoses in domestic and wild ruminants. She obtained a specialty "Parasitology and invasive diseases of animals and humans". Her main scientific interests are parasitology, host-parasite interactions, pulmonary helminthoses, parasitic zoonoses and wildlife parasitoses. Thanks to her many years work as a lecturer in parasitology and wildlife diseases at the Faculty of Veterinary Medicine at the University of Forestry in Sofia, she has extensive academic experience in professional education and training. Dr Panayotova-Pencheva has also proven herself as a project leader and team member, having led and participated in numerous national and international research projects funded by both Bulgarian and foreign sources. Her active involvement in professional networks, including the European and World Parasitology Associations and the Administrative Board of the Bulgarian Society of Parasitology, underlines her role in shaping scientific collaboration and policy dialogue in the field. She is a member of the editorial and review boards of several scientific journals in the field of parasitology and veterinary medicine. Dr Panayotova-Pencheva has authored over 150 scientific publications and regularly participates in scientific forums and conferences to further develop her scientific and professional skills.

Preface

Parasitism is a widespread biological phenomenon and is an inseparable part of life in zoological gardens. The negative effects of parasite infestation in zoo animals can be: the development of secondary deficiencies and other infections; impaired reproduction and a risk of death in cases of massive and dangerous parasitoses. Zoos are also places where wild animals come into close contact with humans. This significantly increases the risk of parasitic zoonoses spreading, which poses a threat to the health of the animals themselves, to the zoo staff and to visitors. The control of parasitoses in zoos is subject to various challenges, such as inadequacies in prevention programs, untimely or inaccurate diagnoses, the introduction of parasites via intermediate hosts, paratenic hosts or mechanical vectors (snails, ants, cockroaches, worms, rodents, etc.), persistent infections (e.g. soil-transmitted helminthoses), the need to adapt antiparasitic treatment to a specific species, the development of drug resistance, etc. Success in this battle requires the combined efforts of all zoo professionals: managers, curators, keepers and veterinarians. Scientific advances are also important, though sometimes the courses of action that can be taken are unclear. The Special Issue titled "The Long-Standing Problem of Parasitic Diseases in Zoo Animals: Current Challenges and Searching for Solutions" was created, aiming to collect scientific articles on the topic. We hope that the developments presented will contribute both to the fundamentals of parasitology, to management practices in zoos, and to a better future for animals in captivity.

Mariana Panayotova-Pencheva
Guest Editor

Journal of
Zoological and Botanical Gardens

MDPI

Case Report

Severe Parasite Co-Infection in a Captive Bactrian Camel: Case Report

Mariana Panayotova-Pencheva [1,]* and Francisco Ponce-Gordo [2]

1 Institute of Experimental Morphology, Pathology and Anthropology with Museum, Bulgarian Academy of Sciences, Academician Georgi Bonchev Str., BI. 25, 1113 Sofia, Bulgaria
2 Department of Microbiology and Parasitology, Faculty of Pharmacy, Complutense University, Plaza Ramón y Cajal s/n, 28040 Madrid, Spain; pponce@ucm.es
* Correspondence: marianasp@abv.bg

Abstract: The aim of this study was to document a case of parasite co-infection in a captive Bactrian camel and to supply morphometric data of the found pathogens. It concerned a 20-year-old male animal inhabiting Sofia Zoo, Bulgaria. A decreased appetite and gastrointestinal disorders were observed in it during the summer of 2022. Improvement in the animal's condition was achieved after the administration of antibiotics, sulfonamides, and other symptomatic medicines. However, two weeks after treatment, clinical symptoms reappeared. Then, a diarrheal fecal sample from the animal was subjected to parasitological examination by direct smear and flotation and sedimentation techniques. Multiple infections by helminths (*Trichostrongylus* sp., *Haemonchus* sp., *Oesophagostomum* sp., *Trichuris* sp., and *Dicrocoelium* sp.), ciliates (*Buxtonella cameli*), and protozoa (*Eimeria cameli*) were found, with *E. cameli* being reported for the first time in zoo conditions. Deworming led to the recovery of the general condition and appearance of the animal's feces, but two weeks later, it died suddenly. We considered that the parasitic infection was not the direct cause of the fatal outcome, and its presence, other health disorders, and the advanced age of the animal were among the contributing factors. This case reveals the need to combine planned preventive deworming with routine parasitological diagnostics to take timely and targeted actions to protect the health of animals inhabiting zoo facilities.

Keywords: *Camelus bactrianus*; gastrointestinal parasites; helminths; protozoa; ciliates; parasite morphometry

1. Introduction

The one-humped camel, or dromedary camel (*Camelus dromedaries* Linnaeus, 1758), and the two-humped camel, or Bactrian camel (*Camelus bactrianus* Linnaeus, 1758) (Artiodactyla, Camelidae), are even-toed ungulates distributed mainly in North Africa and Asia, with the majority of their populations domesticated and raised as livestock. A small number of two-humped camels are left in the wild, standing out as a separate species (*Camelus ferus* Przew, 1878) that is critically endangered [1]. Camels are very hardy animals, well adapted anatomically and physiologically to harsh climatic conditions; nevertheless, they suffer from various parasitic diseases, which are major constraints to the improvement of their health [2].

Parasite infections in camels can lead to nutritional and immune inadequacy, stunted growth and delayed development, infertility problems, adverse effects on the quality of meat and milk, reduced working efficiency, and sometimes death [3,4]. The parasite influence increases significantly when the animals are kept at zoos and other closed facilities, where various factors cause wider distribution and difficult control of parasitoses [5]. In recent years, a variety of helminths, protozoa, and ciliates have been reported parasitizing livestock and zoo camels: *Trichostrongylus* spp., *Trichuris* spp., *Ascaris* spp., *Moniezia* spp., and *Eimeria* spp. in Egypt [3,6]; *Ostertagia* spp., *Trichostrongylus* spp., *Haemonchus contortus*, *Nematodirus* spp., *Marshallagia* spp., *Trichuris* spp., *Chabertia ovina*, *Bunostomum* spp., *Strongyloides papillosus*, *Thysaniezia ovilla*, *Moniezia expansa*, *Dicrocoelium* spp., *Fasciola hepatica*, *Hasstilesia ovis*, and *Eimeria* spp. in China [4]; *Nematodirus* spp., *Trichostrongylus* spp., *Haemonchus* spp., *Trichuris* spp. *Marshallagia* spp., and *Eimeria cameli* in Iran [7], *Eimeria* sp. and *Cystoisospora* sp. in different parts of the world [8]; *Paramphistomum* sp., *Fasciola* sp., *Moniezia* sp., *Dicrocoelim* sp., *E. cameli*, *E. dromedarii*, *E. rajasthani*, *E. pellerdyi*, *Cryptospridium* sp., and *Balantidium coli* in Egypt [9]; *E. cameli*, *E. rajasthani*, and *E. pellerdyi* in Saudi Arabia [10]; *Trichuris* sp., Strongylidae, *Eimeria* sp., and *Buxtonella* sp. in the Ljubljana Zoo, Slovenia [11]; *Trichuris* sp. in Zoological Garden "Ogród Zoologiczny" in Warsaw, Poland [12]; *Trichostrongylus* sp., *Cooperia* sp., *Eimeria bactriani*, and *E. dromedarii* in two zoos located in southern Poland [13]; and *Trichuris* spp. in the Bioparco Zoological Garden of Rome, Italy [14].

Timely and accurate identification of parasites in zoos and breeding centers for rare animals can be decisive for the health of animals by helping with the parasite source recognition and its elimination [11,15]. Lifetime diagnosis of parasitic diseases in veterinary practice is most often performed through coprological investigations. Its correctness depends on the accuracy of the techniques and methods used and especially on the morphological identification of the parasite forms. In connection with the above, the purpose of this report was set, namely to document a case of parasite co-infection in a captive Bactrian camel and to supply morphometric data of the found pathogens.

2. Materials and Methods

2.1. Case History

It concerns a 20-year-old male Bactrian camel inhabiting Sofia Zoo (Bulgaria). The animal lived with Cameroonian goats in the same enclosure, which consisted of an outdoor part (main one, covered by soil and vegetation) and an indoor part (covered by concrete). All animals from the zoo, including this camel, were routinely dewormed according to the accepted preventive plan.

In the summer of 2022, the animal's health worsened—profuse diarrhea, lack of appetite, and weight loss were observed. During 20 days, several courses of treatment were carried out, in which antibiotics, sulfonamides, symptomatic antidiarrheal agents, and stabilizers of the normal intestinal flora were used. As a result of the treatment, the condition of the animal normalized, but two weeks later, diarrhea reappeared. Then, zoo officials approached us with a request to carry out a parasitological examination of the camel.

2.2. Laboratory Investigations

Fresh diarrheal feces were received in the laboratory. A part of the sample was examined immediately, and another part was stored at room temperature (22–25 °C) and was examined on the 10th and 17th days after its receipt. Microscopic examinations were carried out by common flotation (salt solution gravity = 1.18) and sedimentation techniques and direct smear [8]. Imaging and measurement of parasite forms were performed using a Motic Images Plus 3.0 camera connected to an Amplival microscope with accompanying software. Parasite forms were identified morphologically according to the descriptions by Kotelnikov [16], Yagoub [17], Thienpont et al. [18], Foreyt [19], and Abbas et al. [6]. The metric data were statistically analyzed using Microsoft® Excel software—version number 2013.

3. Results

Microscopic examinations of the fresh feces (on the first day after collection) by the flotation technique revealed four types of nematode eggs, three of which were strongylid ones. The strongylid eggs of the first type (Figure 1a) were asymmetrical, with dissimilar poles (one of which was more rounded than the other), dissimilar side walls, smooth shell surface, and numerous blastomeres. The strongylid eggs of the second type (Figure 1b) were a regular ellipse in shape, with nearly similar wide poles, barrel-shaped side walls, smooth shell surface, and numerous hardly distinguishable blastomeres. The third type of strongylid eggs was similar to the second one, but their blastomeres were clearly distinguished (Figure 1c). Nematode eggs of the fourth type were lemon-shaped and brown-colored, with two protruding, transparent polar plugs, thick walls, and granular content (Figure 1d). Considering the morphological criteria for identification (shape, poles, walls, surface, and content) and the metric features (Table 1), the eggs observed were assigned to the *Trichostrongylus* sp. (asymmetrical, dissimilar poles, and 81–94 μm), *Haemonchus* sp. (elliptical, barrel-shaped side walls, hardly distinguishable blastomeres, and 72–90 μm), *Oesophagostomum* sp. (elliptical, clearly distinguished blastomeres, and 76–85 μm), and *Trichuris* sp. (lemon-shaped, two protruding polar plugs, and 62–72 μm), respectively.

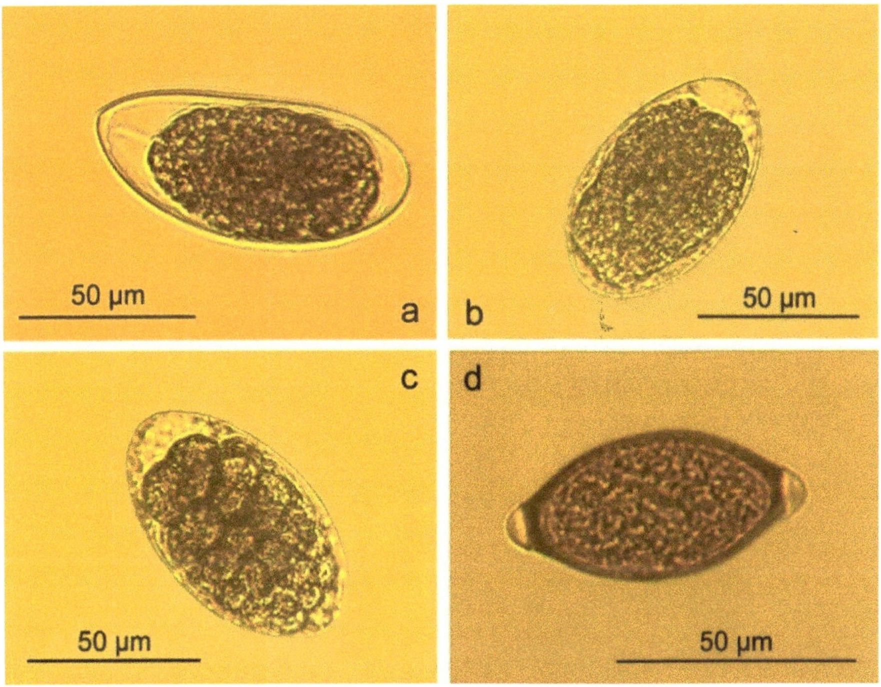

Figure 1. Nematode eggs found in the feces of a *Camelus bactrianus* with multiple parasite infections: (**a**) *Trichostrongylus* sp.; (**b**) *Haemonchus* sp.; (**c**) *Oesophagostomum* sp.; (**d**) *Trichuris* sp. Original pictures.

Examination of feces by the sedimentation technique revealed small (37–40 μm), asymmetrical, and dark-brown, with one operculum and two eye spots, trematode eggs (Figure 2), which were identified as *Dicrocoelium* sp. Oocysts with average sizes of 87/63 μm were also found through the sedimentation technique. They were ovoid shaped and brown to black in color, with a three-layer wall (outer and inner dark brown and middle yellowish) and a micropyle, which was not always visible (Figure 3). They were morphometrically identified as *Eimeria cameli* (Henry and Masson, 1932) Reichenow, 1952.

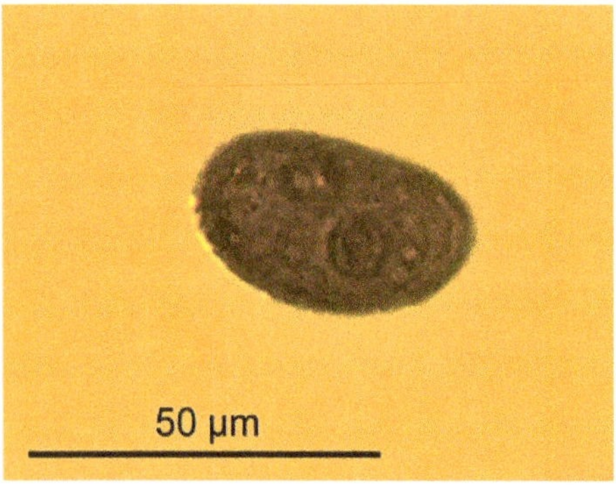

Figure 2. *Dicrocoelium* sp. egg found in the feces of a *Camelus bactrianus* with multiple parasite infection. Original picture.

Figure 3. *Eimeria cameli* oocysts found in the feces of a *Camelus bactrianus* with multiple parasite infections. Original pictures.

Direct smear revealed motile unicellular forms with a morphology (Figure 4a) and dimensions (Table 1) resembling those of trophozoites of *Buxtonella cameli* (Boschenko 1925) Esteban-Sánchez et al. 2023. On the tenth day after fecal collection, active trophozoites were still observed. They were relatively less than those visualized in the fresh feces. Motile formations with the same morphology that were smaller in size and closely contacted side by side were also found (Figure 4b), with this probably being the trophozoites in the process of reproduction.

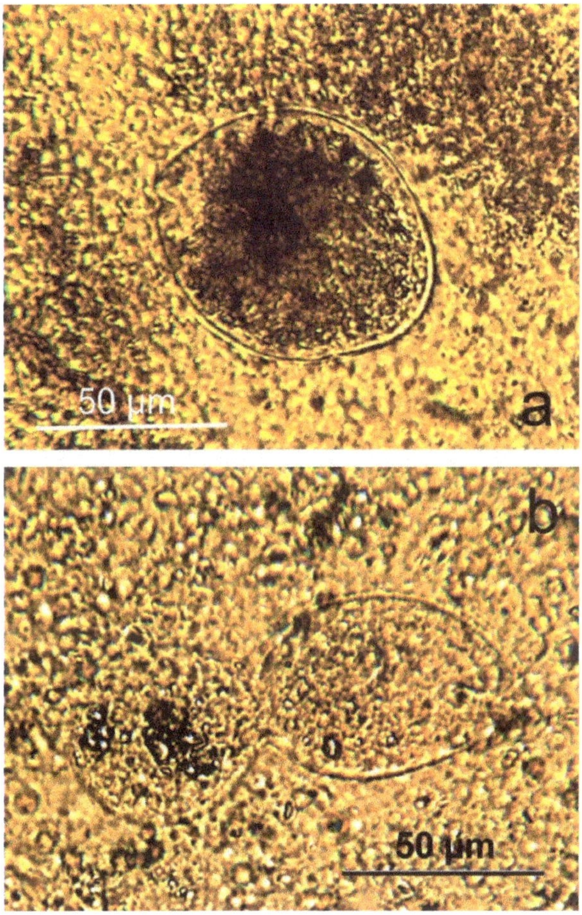

Figure 4. *Buxtonella cameli* trophozoites found in the feces of a *Camelus bactrianus* with multiple parasite infections: (**a**) On the first day after fecal collection. (**b**) On the tenth day after fecal collection. Original pictures.

Gastrointestinal strongylid larvae were found in the diarrheal feces (by direct smear) on day 17 of their collection. They were medium in size (Table 1), with the following morphological features: square head end; tail end with a small spike at the tip; presence of a sheath, which formed a short cone on the caudal larval end; and 16 triangular intestinal cells arranged in two rows, with sharp tips pointing forward and backward (Figure 5). The larvae observed were assigned to *Trichostrongylus* sp.

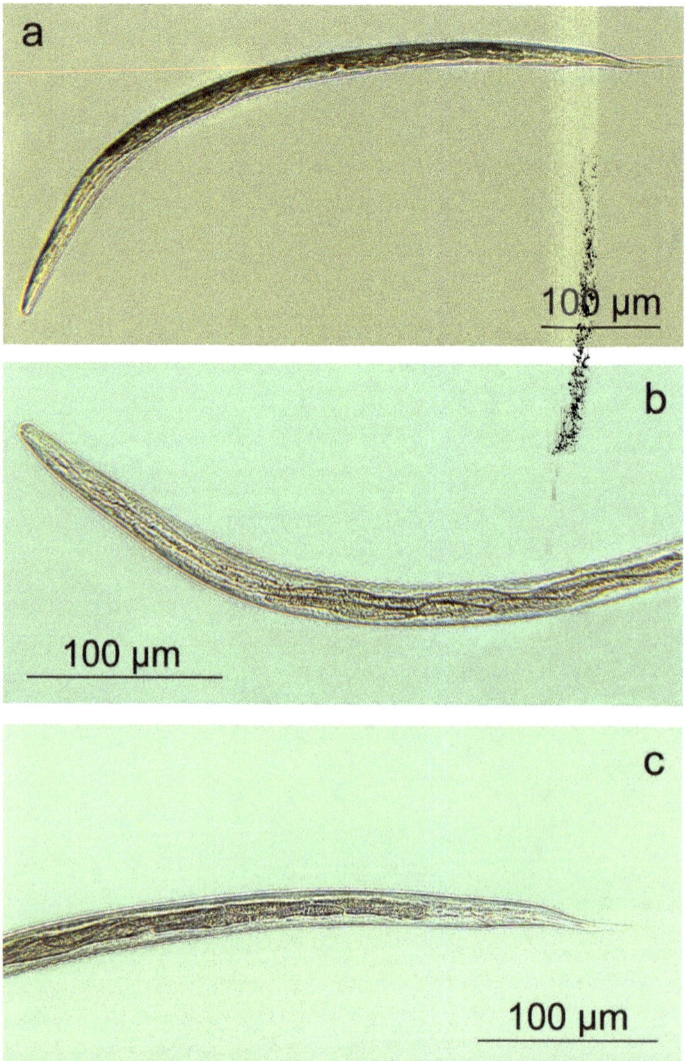

Figure 5. *Trichostrongylus* sp. larva found in the feces of a *Camelus bactrianus* with multiple parasite infections. (**a**) Whole larval body. (**b**) Anterior body end. (**c**) Posterior body end. Original pictures.

Table 1. Comparative measurements (in micrometers) of parasites from a *Camelus bactrianus* (present case) and other sources.

Parasite		Present Case		Other Sources		
		Range	Mean ± SD	Range	Hosts	References
Trichostrongylus sp. eggs	L	81–94	86.18 ± 4.96	70–125	Cattle, Sheep, Goats	Thienpont et al. [18] Foreyt [19]
	W	37–46	42.45 ± 2.62	30–55		
Haemonchus sp. eggs	L	72–90	83 ± 4.94	62–95	Cattle, Sheep, Goats	Thienpont et al. [18] Foreyt [19]
	W	44–50	47.33 ± 1.72	36–50		

Table 1. *Cont.*

Parasite		Present Case		Other Sources		
		Range	Mean ± SD	Range	Hosts	References
Oesophagostomum sp. eggs	L	76–85	81 ± 3.24	65–120	Cattle, Sheep, Goats	Thienpont et al. [18] Foreyt [19]
	W	41–48	44.22 ± 2.05	40–60		
Trichuris sp. eggs	L	62–72	66.8 ± 3.83	70–80	Cattle, Sheep, Goats	Thienpont et al. [18] Foreyt [19]
	W	27–32	30.2 ± 2.17	30–42		
Trichostrongylus sp. L1	L	480–676	602.33 ± 106.68	650–770	Ruminants	Kotelnikov [16]
	W	20–23	21.33 ± 1.53	-		
Dicrocoelium sp. eggs	L	37–40	38.33 ± 1.21	38–45	Cattle, Sheep	Thienpont et al. [18]
	W	23–24	23.33 ± 0.52	22–30		
Eimeria cameli oocysts	L	76–96	86.62 ± 6.05	70–100	Dromedary camels	Abbas et al. [6] Yagoub [17]
	W	56–70	63.10 ± 3.65	52.5–73.8		
Buxtonella cameli trophozoite	L	64–86	74.5 ± 6.29	30–150	Pigs	Foreyt [19]
	W	51–72	61.9 ± 5.15	-		

L1—first-stage larva; L—length; W—width.

Following the initial laboratory studies, the animal was consecutively treated with sulfaguanidine (0.1 g/kg q 24 h × 4 d PO) and fenbendazole (15 mg/kg q 24 h × 5 d PO), considering the recommendations for treating parasitic infections in camels [20]. On the first day after treatment, a fecal sample was obtained for a control test. The feces were fully formed and of normal consistency and color. Eggs of *Dicrocoelium* sp. and single ones of *Trichuris* sp. were observed microscopically. *Eimeria* and ciliates were not detected. Ten days later, the camel died suddenly, and abdominal distension and liver damage were detected. There were no other data from the autopsy, and subsequent parasitological investigations of the dead camel were not performed.

4. Discussion

In this case, we are dealing with a severe infection in a camel infested with numerous parasites from different taxa-nematodes, trematodes, coccidia, and ciliates. According to available research, gastrointestinal parasites are the most frequently registered pathogens in camels, and nematodes of orders Strongylida and Enoplida and protozoa of *Eimeria* sp. are the most common of them [3,4,6,7,9,10,17,21–23]. The present findings of *Trichostrongylus* sp., *Haemonchus* sp., *Oesophagostomum* sp., *Trichuris* sp., and *E. cameli* confirm this trend. Previous studies showed that camels were often co-infected with a large number of parasite species [7,9,24,25], reaching up to 14 different species in a single animal [4]. Having in mind favorable conditions for parasites spreading in zoos, the occurrence of this multiple infection in the present case is not a surprise. Infestation of camels with gastrointestinal strongylids and trichurids were also found in a number of European zoos [11–14,23,26–28], including the Sofia Zoo, where *Trichuris* sp. and *Nematodirus* sp. were registered more than 30 years ago [29].

Literature reports of *E. cameli* establishment are relatively few. This protozoan was found in livestock camels [6,10,17], but according to our best knowledge, there are no records in zoo animals. It should be borne in mind that the diagnosis of *E. cameli* could be missed, especially in captive camels. This could be for several reasons: Infected adult animals in a good general condition rarely excrete oocysts in feces, or excreted oocysts are in low numbers [8]; the excretion of oocysts does not begin immediately with the development of clinical symptoms but occurs later [22]; and false negative results could occur because of technical reasons—*E. cameli* oocysts are large and heavy [8] and may not be detected by flotation coproscopic methods using low-gravity solutions. Probably because of the third reason, we did not observe oocysts by common flotation in the present case. Here, we found *E. cameli* oocysts only by the sedimentation technique. It is necessary to keep in mind that solutions of specific gravity > 1.28 are recommended for the floatation of this protozoan [22].

Trophozoites of the ciliate parasite that we observed in this case initially were thought to correspond to *Balantioides coli* [(Malmsten 1857) Stein 1863] Alexeieff 1931, as it is the most frequently cited ciliate (as *Balantidium coli*) in camels [9,30–33]. However, *Buxtonella* sp. has been also cited in camels [11,34]. Moreover, one month after the death of the camel, we received fecal samples from other healthy camels of the Sofia Zoo with a request for a prophylactic parasitological examination. These camels entered the zoo a year ago and lived separately from the dead camel. Parasitological tests of the new camels revealed ciliate cysts that were identified by genetic analysis as *B. cameli* [35]. This led us to consider that the ciliates found in the deceased camel likely corresponded to *B. cameli*.

The severe progression of the infection was influenced not only by its multiplicity but also by certain factors associated with zookeeping. The camel lived in an area with a natural covering of grass and bushes, which was difficult to disinfect and clear of parasites. A study at Dublin Zoo, for example, showed that strongylid eggs and larvae persist in soil and grass of camel enclosures, with 1500 *Trichostrongylus* larvae found in 1 kg of grass [36]. The joint keeping of the camel with Cameroonian goats probably contributed to the increased parasite load on the environment and hence the animals. Such a positive correlation between cross-species animal contact and parasite infestation was found in two zoos in Poland, where the level of parasite infection was higher in camels kept in the same enclosure with alpacas and Shetland ponies than those kept separately [13].

Clinical symptoms, such as anorexia, weight loss, and diarrhea observed in the present case, are usual in camels with gastrointestinal parasite infestation [30,37–39]. However, the treatment of such infections has yielded varied results. For example, severe trichurid infection was successfully cured in young 3-year-old camels, but regardless of the same treatment, it ended fatally in an adult 13-year-old animal [37]. In the present case, the advanced age of the animal and the concurrent infestation, factors determining the severe course of gastrointestinal infections [40], probably contributed to the fatal outcome, despite the treatment undertaken.

Morphometric features of parasites found, as a whole, corresponded to those from other sources (Table 1). Only the dimensions of *Trichuris* sp. eggs and first-stage larvae of *Trichostrongylus* sp. were smaller or closer to the lower limits than indicated in manuals [16,18,19]. Certain varieties in color and morphological structures of *E. cameli* oocysts were shown [9,17]. Some authors even distinguished four different oocyst morphotypes within the species [6]. In this case, we observed oocysts corresponding to the first and fourth morphotypes described by Abbas et al. [6]. Thus, our data confirmed the trend of the morphological diversity of *E. cameli* oocysts; furthermore, they were collected from only one animal.

5. Conclusions

Despite the high endurance of camels in adverse conditions, the current case of multiple parasite infestation in a camel was severe. Keeping the animal in a zoo, together with other herbivores, probably contributed to the complication of the infection. The parasitic infection was probably not a direct cause of the camel's death, but its presence, other health disorders, and advanced age were among the factors that led to the fatal outcome of the animal. This case reveals the need to combine preventive planned deworming in zoos with routine parasitological diagnostics in order to take timely and targeted actions to protect the health of the animals.

Author Contributions: M.P.-P.: conceptualization, resources, methodology, formal analysis, validation, data curation, writing—original draft, and review and editing. F.P.-G.: formal analysis, validation, and writing—review and editing. All authors have read and agreed to the published version of the manuscript.

Funding: This research received no external funding. It was performed on a voluntary basis, according to an agreement for collaboration between the Institute of Experimental Morphology, Pathology, and Anthropology with Museum-Bulgarian Academy of Sciences and the Sofia Zoo.

Institutional Review Board Statement: Not applicable.

Data Availability Statement: All data is provided within the manuscript.

Conflicts of Interest: The authors declare no conflict of interest.

References

1. Hare, J. *Camelus ferus* Przewalski, 1878. The IUCN Red List of Threatened Species. 2008. Available online: https://www.iucnredlist.org/species/63543/12689285 (accessed on 6 October 2023).
2. Parsani, H.R.; Singh, V.; Momin, R.R. Common parasitic diseases of camel. *Vet. World* **2008**, *1*, 317–318.
3. Hamed, M.I. Ivermectin resistance in intestinal parasites of camels in a private farm at Assiut, Egypt. *Comp. Clin. Path.* **2018**, *27*, 1221–1226. [CrossRef]
4. Guowu, Z.; Kai, Z.; Xifeng, W.; Chunhui, J.; Chengcheng, N.; Yue, Z.; Jun, Q.; Qingling, M.; Xingxing, Z.; Kuojun, C.; et al. Occurrence of gastrointestinal parasites in camels in the Tianshan Mountains pastoral area in China. *J. Vet. Res.* **2020**, *64*, 509–515. [CrossRef] [PubMed]
5. Panayotova-Pencheva, M.S. Experience in the Ivermectin treatment of internal parasites in zoo and captive wild animals: A review. *Der Zoolog. Garten* **2016**, *85*, 280–308. [CrossRef]
6. Abbas, I.E.; El-Alfy, E.; Al-Araby, M.; Al-Kappany, Y.; El-Seadaw, R.; Dubey, J.P. Prevalence of *Eimeria* species in camels (*Camelus dromedarius*) from Egypt and variability in structure of *Eimeria cameli* oocysts. *J. Parasitol.* **2019**, *105*, 395–400. [CrossRef]
7. Radfar, M.H.; Gowhari, M.A. Common gastrointestinal parasites of indigenous camels (*Camelus dromedarius*) with traditional husbandry management (free-ranging system) in central deserts of Iran. *J. Parasit. Dis.* **2013**, *37*, 225–230. [CrossRef]
8. Dubey, J.P.; Schuster, R.K. A review of coccidiosis in Old World camels. *Vet. Parasitol.* **2018**, *262*, 75–83. [CrossRef]
9. El-Khabaz, K.A.S.; Abdel-Hakeem, S.S.; Arfa, M.I. Protozoan and helminthes parasites endorsed by imported camels (*Camel dromedaries*) to Egypt. *J. Parasitol. Dis.* **2019**, *43*, 607–615. [CrossRef]
10. Metwally, D.M.; Al-Otaibi, T.T.; Albasyouni, S.A.; El-Khadragy, M.F.; Alajmi, R.A. Prevalence of eimeriosis in the one-humped camels (*Camelus dromedarius*) from Riyadh and Al-Qassim, Saudi Arabia. *Peer J.* **2020**, *8*, e10347. [CrossRef]
11. Kvapil, P.; Kastelic, M.; Dovč, A.; Bártová, E.; Čížek, P.; Lima, N.; Štrus, Š. An eight-year survey of the intestinal parasites of carnivores, hoofed mammals, primates, ratites and reptiles in the Ljubljana zoo in Slovenia. *Folia Parasitol.* **2017**, *64*, 013. [CrossRef]
12. Maesano, G.; Capasso, M.; Ianniello, D.; Cringoli, G.; Rinaldi, L. Parasitic infections detected by flotac in zoo mammals from Warsaw, Poland. *Acta Parasitol.* **2014**, *59*, 343–353. [CrossRef] [PubMed]
13. Nosal, P.; Kowal, J.; Kornaś, S.; Wyrobisz, A.; Skotnicki, J.; Basiaga, M.; Plucińska, N.E. Endoparasites of exotic ungulates from the Giraffidae and Camelidae families kept ex situ. *Ann. Parasitol.* **2016**, *62*, 67–70. [CrossRef] [PubMed]
14. Di Filippo, M.M.; Berrilli, F.; De Liberato, C.; Di Giovanni, V.; D'Amelio, S.; Friedrich, K.G.; Cavallero, S. Molecular characterization of *Trichuris* spp. from captive animals based on mitochondrial markers. *Parasitol. Int.* **2020**, *75*, 102043. [CrossRef] [PubMed]
15. Eleni, C.; Di Cesare, A.; Cavicchio, P.; Tonnicchia, M.C.; Meoli, R.; di Regalbono, A.F.; Paoletti, B.; Pietrobelli, M.; De Liberato, C. Fatal *Angiostrongylus dujardini* infection in callitrichid monkeys and suricates in an Italian zoological garden. *Parasitol. Int.* **2016**, *65*, 333–335. [CrossRef]
16. Kotelnikov, G.A. *Gelminthological Animal and Environment Investigation*; Kolos: Moscow, Russia, 1984; p. 208. (In Russian)
17. Yagoub, I.A. Coccidiosis in Sudanese camels (*Camelus dromedarius*): 1-First record and description of *Eimeria* spp. Harboured by Camels in the Eastern Region of Sudan. *J. Protozool.* **1989**, *36*, 422–423. [CrossRef] [PubMed]
18. Thienpont, D.; Rochette, F.; Vanparijs, O. *Diagnosing Helminthiasis through Coprological Examination*, 3rd ed.; Jansen Animal Health: Beerse, Belgium, 2003; p. 215.
19. Foreyt, W.J. *Veterinary Parasitology Reference Manual*, 5th ed.; Blackwell Publishing: Hoboken, NJ, USA; Iowa State University Press: Ames, IA, USA, 2001; p. 235.
20. Walker, P.G. Camelid Medicine Cabinet. Available online: https://icinfo.vet.ohio-state.edu/sites/camelid-sta.osumc.edu/files/documents/Camelid_Medicine_Cabinet.pdf (accessed on 6 October 2023).
21. Banaja, A.A.; Guandour, A.M. A review of parasites of camels (*Camelus dromedarius*) in Saudi Arabia. *Science* **1994**, *6*, 75–86. [CrossRef]
22. Dubey, J.P. A review of coccidiosis in South American camelids. *Parasitol. Res.* **2018**, *117*, 1999–2013. [CrossRef]
23. Zanzani, S.A.; Villa, L.; Gazzonis, A.L.; Cartagena, D.; Mortarino, M.; Bonacina, E.; Guadagnini, D.; Allievi, C.; Manfredi, M.T. The prophylactic effect of ivermectin treatments on nematode infections of mammals in a Faunistic Park (Northern Italy). *Animals* **2022**, *12*, 1124. [CrossRef]
24. Osman, F.; Mustafa, H.; Sayed, G. Clinico- hematological and biochemical changes in camels infected with gastro-intestinal parasites. *J. Anim. Sci. Adv.* **2015**, *5*, 1245–1252. [CrossRef]

25. Bekele, J.T.; Aregawi, W.G.; Wegi, F.G.; Geletu, A.S.; Tesfamariam, W. Epidemiological investigation of gastrointestinal parasites of dromedary camels in Administrative Zone Three of Afar Region, Ethiopia. *Vet. Med. Int.* **2022**, *2022*, 8433997. [CrossRef]
26. Geraghty, V.; Mooney, J.; Pike, K. A study of parasitic infections in mammals and birds at the Dublin Zoological Gardens. *Vet. Res. Commtm.* **1982**, *5*, 343–348. [CrossRef] [PubMed]
27. Fagiolini, M.; Lia, R.P.; Laricchiuta, P.; Cavicchio, P.; Mannella, R.; Cafarchia, C.; Otranto, D.; Finotello, R.; Perrucci, S. Gastrointestinal parasites in mammals of two Italian zoological gardens. *J. Zoo Wildl. Med.* **2010**, *41*, 662–670. [CrossRef] [PubMed]
28. Atanaskova, E.; Kochevski, Z.; Stefanovska, J.; Nikolovski, G. Endoparasites in wild animals at the zoological garden in Skopje, Macedonia. *J. Threat. Taxa* **2011**, *3*, 1955–1958. [CrossRef]
29. Matevsky, S. Helminths and helminthoses in zoo animals in Sofia. *Verh. Erkrg. Zootiere* **1988**, *30*, 173–176. (In Russian)
30. Tajik, J.; Nourollahi Fard, S.R.; Paidar, A.; Anousheh, S.; Dehghani, E. Balantidiasis in a dromedarian camel. *Asian Pac. J. Trop. Dis.* **2013**, *3*, 409–412. [CrossRef]
31. Hussein, A.G.; Al-Samarai, F.R.; Asal, S.N. Impact of age and gender of camels (*Camelus dromedarius*) and their breeders on the prevalence of *Balantiium coli*. *J. Biosci. Agric. Res.* **2016**, *8*, 734–738. [CrossRef]
32. Osman, S.A. Balantidiasis in dromedary camels: Prevalence, haematology and treatment outcomes. *J. Camel Pract. Res.* **2019**, *26*, 81–85. [CrossRef]
33. Saidi, R.; Mimoune, N.; Chaibi, R.; Baazizi, R.; Abdelouahed, K.; Khelef, D.; Kaidi, R. Camel gastrointestinal parasites in southern Algeria. *Vet. Stanica* **2022**, *53*, 283–294. [CrossRef]
34. Bouragba, M.; Laatamna, A.K.; Cheddad, F.E.; Baroudi, D.; Houali, K.; Hakem, A. Gastrointestinal parasites of dromedary camel (*Camelus dromedarius*) in Algeria. *Vet. World* **2020**, *13*, 1635–1640. [CrossRef]
35. Esteban-Sanchez, L.; Panayotova-Pencheva, M.; Qablan, M.; Modrý, D.; Hofmannova, L.; Ponce-Gordo, F. Question of agent of camel balantidiosis solved: Molecular identity, taxonomic solution and epidemiological considerations. *Vet. Parasitol.* **2023**, *321*, 109984. [CrossRef]
36. Scullion, F.T. An investigation into the parasite burden available to Grant's zebras (*Hippotigris burchelli*) and Bactrian camels (*Camelus bactrianus*) in the Dublin Zoo. *J. Zoo An. Med.* **1982**, *13*, 156–160. [CrossRef]
37. Eo, K.Y.; Kwak, D.; Kwon, O.D. Severe whipworm (*Trichuris* spp.) infection in the dromedary (*Camelus dromedarius*). *J. Zoo Wildl. Med.* **2014**, *45*, 190–192. [CrossRef] [PubMed]
38. AL-Tayib, O. Zoonotic balantidiasis in camel from Saudi Arabia. *Sch. Acad. J. Biosci.* **2014**, *2*, 445–447.
39. Locklear, T.R.; Videla, R.; Breuer, R.M.; Mulon, P.-Y.; Passmore, M.; Mochel, J.P.; Gerhold, R.; Schaefer, J.J.; Smith, J.S. Presentation, clinical pathology abnormalities, and identification of gastrointestinal parasites in camels (*Camelus bactrianus* and *Camelus dromedarius*) presenting to two North American veterinary teaching hospitals. A Retrospective Study: 1980–2020. *Front. Vet. Sci.* **2021**, *8*, 651672. [CrossRef]
40. Paul, B.T.; Jesse, F.F.A.; Chung, E.L.T.; Che'Amat, A.; Mohd Lila, M.A. Risk factors and severity of gastrointestinal parasites in selected small ruminants from Malaysia. *Vet. Sci.* **2020**, *7*, 208. [CrossRef] [PubMed]

Journal of
*Zoological and
Botanical Gardens*

MDPI

Article

Respiratory Infection by *Cyathostoma (Hovorkonema) americana* in a Population of Burrowing Owls (*Athene cunicularia*)—A Potential Case of Zoo–Wildlife Cross-Transmission

Gonçalo N. Marques [1,†], João T. Cruz [2,3,*,†], Matilde Pinto [2,3], Miriam Leal [1], Carla Flanagan [1], Nuno Urbani [1] and Luís Madeira de Carvalho [2,3]

[1] Zoomarine, 8201-864 Guia, Portugal; goncalo.marques@zoomarine.pt (G.N.M.); miriam.leal@zoomarine.pt (M.L.); carla.flanagan@zoomarine.pt (C.F.); nuno.urbani@zoomarine.pt (N.U.)
[2] CIISA—Centre for Interdisciplinary Research in Animal Health, Faculty of Veterinary Medicine, University of Lisbon, 1300-477 Lisbon, Portugal; matilde.goes@campus.ul.pt (M.P.); madeiradecarvalho@fmv.ulisboa.pt (L.M.d.C.)
[3] Associate Laboratory for Animal and Veterinary Sciences (AL4AnimalS), 1300-477 Lisbon, Portugal
* Correspondence: joaot99fmv@gmail.com
† These authors contributed equally to this work.

Abstract: A population of burrowing owls (*Athene cunicularia*) under professional care at *Zoomarine* Portugal presented with sudden respiratory clinical signs. Clinical management included a thorough diagnosis plan, including in-house fecal analysis that revealed the presence of ovoid unioperculate eggs. In the *postmortem* examination of one hyperacute dyspneic specimen, adult nematode parasites were collected and identified based on their morphology as *Cyathostoma (Hovorkonema) americana.* Even after a broad-spectrum deworming protocol as part of the treatment and metaphylaxis approach, the incidence of parasitic reinfection was high. The complete clinical resolution was only accomplished after the identification and management of the possible focus of infection, a wild population of cattle egrets (*Bubulcus ibis*) that frequently congregated above the owls' habitat. To the authors' best knowledge, this is the first report of infection by *Cyathostoma* (*Hovorkonema*) *americana* in burrowing owls. Although nematodes of the family Syngamidae are not commonly included in the differential diagnosis of infectious respiratory agents of birds of the order Strigiformes, this report highlights the possibility of opportunistic parasitism in a zoological context, especially where there is a continued proximity to free-ranging avifauna.

Keywords: *Athene cunicularia*; *Cyathostoma* (*Hovorkonema*) *americana*; respiratory disease; opportunistic parasitism; zoo–wildlife cross-transmission

1. Introduction

Burrowing owls (*Athene cunicularia*) have a wide distribution on the continents of North and South America, and their conservation status is defined by the International Union of Conservation of Nature (IUCN) as of least concern, although there is a decreasing population trend [1,2]. Given the important role of these birds in maintaining a balance in the populations of their prey and also serving as prey themselves for other birds of prey, the decline of some populations may be causing a ripple effect throughout the ecosystem [3,4]. Due to seasonal changes on the population density of their prey, burrowing owls have adaptable food habits. Approximately 90% of burrowing owls' diet is based on arthropods, including grasshoppers, crickets, and beetles, although they also eat small mammals, such as mice, small birds, and ground squirrels [3,5,6]. The main threat to these owls is the loss of habitat, mainly due to human activity [2–4]. Burrowing owls' habitats consist of open-canopied areas, with sparse ground vegetation. These habitats may include agricultural lands, grasslands, prairies, plains, and deserts [3,4].

These small birds of prey have diurnal habits and use burrows in the ground for refuge [3]. They may dig their own burrows or use others abandoned by other animals such as prairie dogs, ground squirrels, badgers, tortoises, coyotes, and foxes. These burrows may reach more than 3 m in length, angled downwards [3,4].

Burrowing owls live 6–8 years in the wild and up to 10 years under professional care in captivity [3,7]. In the wild, two-thirds do not live to adulthood, with mortality frequently happening between fledging and the end of first year of age. The causes of mortality at a young age include low prey density, inexperience in capturing food, predators, and parasitic infections [3].

This species of owl is found in several zoological institutions. A burrowing owl population of 11 resident individuals at *Zoomarine* Portugal inhabited an outdoor walkthrough enclosure adapted for a variety of species of birds and reptiles. As part of the park's zoological collection, these individuals were included in a preventative medicine program and had no relevant clinical history until the sudden and sequential development of respiratory clinical signs.

The subfamily Syngaminae Baylis & Daubney, 1926 includes parasitic nematode organisms found in the respiratory systems of both avian and mammalian hosts, with representatives such as *Boydinema*, *Cyathostoma*, and *Syngamus* for birds and *Mammomonogamus* and *Rodentogamus* for mammals. Notably, the genus *Cyathostoma* was established by Blanchard in 1849 and stands out as the most extensive, boasting a population exceeding 20 distinct species [8]. *Cyathostoma* sp. were reported in several bird species and, even though the occurrence of these nematodes is generally subclinical, heavy infections are often associated with secondary infections and death [8–14]. The ecology of *Cyathostoma* sp. parasites is important for veterinary and conservation reasons and is still a challenging taxonomy and a contentious topic [8].

In the most recent system of Syngaminae proposed by Lichtenfels in 1980 [15], despite being more than 40 years old, the genus *Cyathostoma* is divided into two subgenera based on the structure of the copulatory bursa and spiculae length. On the one hand, *Cyathostoma* (*Cyathostoma*) (Blanchard, 1849) exhibits a dorsal ray that extends beyond the end of the copulatory bursa, forming characteristic thorn-like projections and spicules that measure 0.08–0.4 mm. On the other hand, *Cyathostoma* (*Hovorkonema*) Turemuratov, 1963 features a dorsal ray that does not extend beyond the end of the copulatory bursa, and spicules are within the range of 0.45–0.8 mm.

Cyathostoma (*Hovorkonema*) *americana* was first described by Chapin in 1925, based on material collected from a red-tailed hawk (*Buteo jamaicensis* Gmelin, 1788) in Virginia (USA), and, while there are occasional subsequent reports of this species in birds of prey [8,10,11,13,14,16–18], there are no available reports of C. (*Hovorkonema*) *americana* in *A. cunicularia.*

This study aims to describe the diagnostic approach for the respiratory infection in a population of *A. cunicularia* held in a zoo, primarily focusing on parasite identification, exploring the evidence of cross-transmission, and investigating opportunistic parasite infections within the unique environment of a zoological setting.

2. Materials and Methods

The burrowing owls with respiratory clinical signs went through a multimodal diagnostic approach, which included a full physical examination under general anesthesia with isoflurane, when possible, according to the clinical status of each individual. Blood sampling was performed for an in-house complete blood count and general biochemistry panel, the latter through a VETSCAN VS chemistry analyzer (Avian/Reptilian Profile Plus). Both total leukocyte and erythrocyte counts were manually completed, in which a 5 μL blood-filled pipette was inserted into a Natt-Herricks-TIC (Bioanalytic GmbH, Umkirch/Freiburg, Germany) 1:200 stain solution vial, and counting was performed with a Neubauer chamber. Evaluation of blood smears was performed after Diff-Quik staining. Hemoglobin levels were obtained through a hemoglobin analyzer HemoCue (HemoCue AB, Ängelholm, Sweden) and hematocrit after centrifugation of microhematocrit tubes (Centurion Scientific

Ltd.—Pro-Vet, West Sussex, UK) at 12,000 rpm for 5 min. Radiographic studies included ventrodorsal and left lateral views with portable radiographic equipment (GIERTH HF300, GIERTH X-RAY International GmbH, Riesa, Germany). For in-house fecal wet mounts, one drop of new methylene blue was added immediately before microscopic observation.

The adult nematode parasites were sent to the Faculty of Veterinary Medicine at the University of Lisbon for analysis conducted by the Parasitology and Parasitic Diseases Service. The parasites were subjected to morphological identification and measurement and underwent the following procedures: (i) examination under a stereomicroscope for initial observation; (ii) mounting on slides with Hoyer's medium for subsequent examination using a compound microscope.

The identification of parasite subgenera employed the Syngaminae classification system as proposed by Lichtenfels [15]. For the identification of the parasite species, the authors consulted the review of *Cyathostoma* sp. nematodes by Kanarek et al. [8]. Detailed observations and measurements encompassed body length, body width, depth of buccal capsule, spicule length in males, and egg dimensions.

3. Results

A substantial proportion, specifically 45% (5/11), of the population of burrowing owls developed clinical signs of respiratory origin. These clinical signs included dyspnea (5/5), tachypnea (5/5), crackles (3/5), and sneezes (1/5). Other clinical findings included compression of the abdominal air sacs' radiographic silhouette (3/5), mild leukocytosis (2/5), eosinophilia (1/5), and subcutaneous emphysema (1/5) (Figure 1). Additionally, observation of the direct fecal smears revealed the presence of ovoid unioperculate eggs (Figure 2).

The postmortem examination of one individual who presented hyperacute dyspnea showed severe signs of pulmonary congestion. Adult nematodes were found in the parenchyma of the left lung apex. Notwithstanding the signs of air sacculitis, there were no parasites found in the trachea or air sacs. The male parasites exhibited a total length of 1.2 cm, while the females measured 2.3 cm. Notably, the morphological configuration of the dorsal ray of the copulatory bursa, which does not extend beyond the end (Figure 3), was identified as being under the subgenus *Cyathostoma* (*Hovorkonema*) (Turemuratov, 1963).

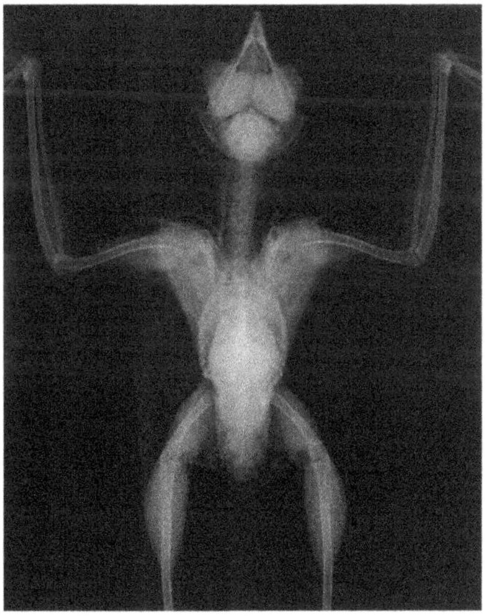

Figure 1. Ventrodorsal radiographic image of an *Athena cunicularia* specimen, depicting a general subcutaneous emphysema.

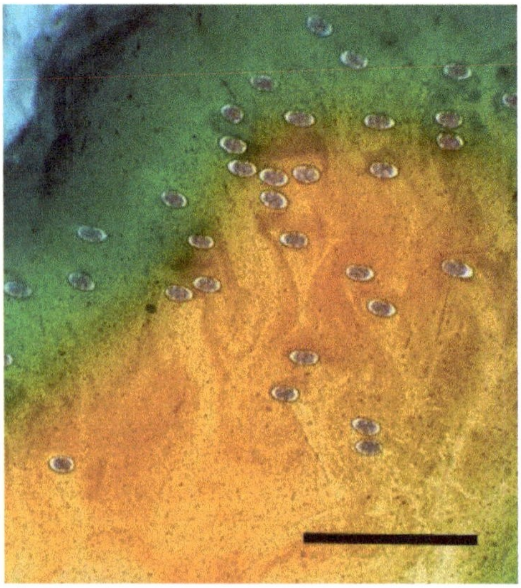

Figure 2. Direct fecal smear showing ovoid unioperculate eggs (new methylene blue, ×100). Bar: 500 μm.

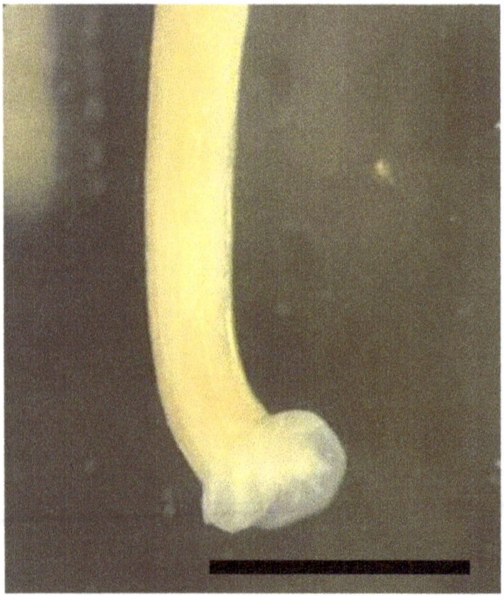

Figure 3. Posterior end of male parasite. Note that no ray of the copulatory bursa extends beyond the end. Bar: 1 mm.

The ensuing measurements are all expressed in micrometers (μm). The dimensions of the male buccal capsule (width × depth) were 255 × 255 (Figure 4), while the female's buccal capsule measured 350 × 210 (Figure 5). The male spicule length measured 435 (Figure 6), and the eggs exhibited dimensions of 85–92.5 in length and 50 in width, with an ovoid form and a single operculum (Figure 7). Notably, given the prevailing attributes, including the smaller spicules and the host being part of the parasite's type group (birds of prey), the specimens were identified as *C.* (*Hovorkonema*) *americana* [16].

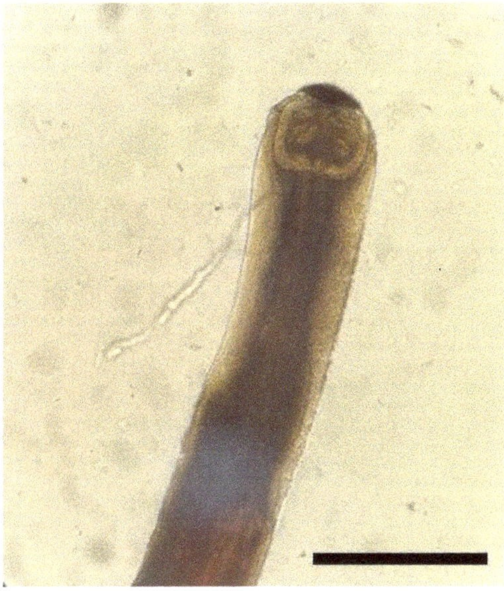

Figure 4. Anterior end of male parasite. Bar: 500 μm.

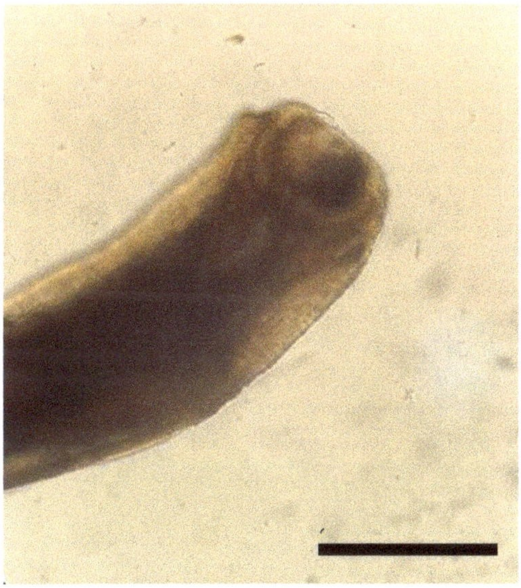

Figure 5. Anterior end of female parasite. Bar: 500 μm.

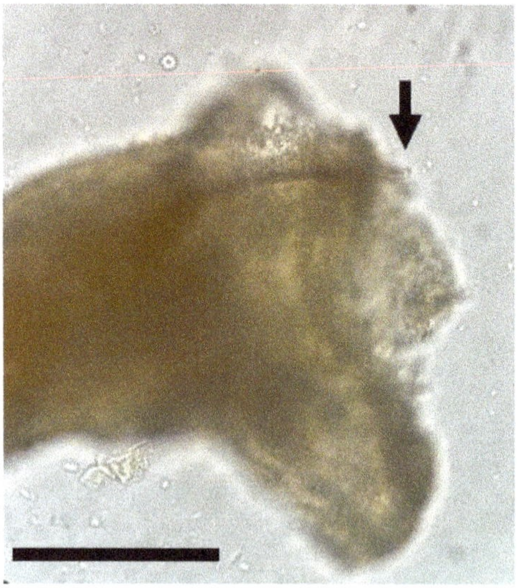

Figure 6. Posterior end of male parasite with copulatory bursa. Note the spicule end (arrow). Bar: 250 µm.

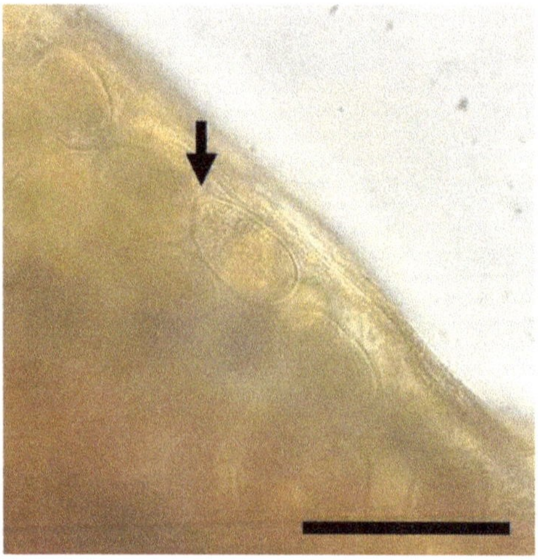

Figure 7. Detail of parasite eggs inside a female specimen. Note the operculum (arrow). Bar: 120 µm.

4. Discussion

Animal welfare assessment is crucial in modern zoological institutions in order to assure the highest possible standards of welfare for animals under professional care. This assessment is based on the Five Domains model, which includes nutrition, environment, health, behavior, and mental state [19–21]. The negative effects of parasitic infections in zoos are numerous and may extensively affect the health dominion (i.e., the absence of diseases, the absence of injuries, and the absence of pain induced by management procedures). The parasitic consequences may vary from the development of secondary infections or nutritional deficiencies to death in the case of severe parasitosis. Moreover, the systematic impact of parasitism may include reproduction impairment, which may be decisive not only for the success of a specific animal collection but also for conservation purposes. Lastly, in a zoological context, close contact between animals and humans is possible, hence the higher risk of the dispersion and spread of parasitic zoonoses [22,23].

The overall management of parasitic infections in zoos may present itself as a challenge, not only from a diagnostic and treatment perspective but also from an environmental point of view. The original population of burrowing owls under professional care at *Zoomarine* was captive-bred in other European institutions and arrived at *Zoomarine* around six years before the onset of the respiratory parasitic infection. Since their arrival, they inhabited a mixed-species outdoor walkthrough enclosure, which included other bird species (giant wood rail (*Aramides ypecaha*), cattle egret (*Bubulcus ibis*), striated heron (*Butorides striata*), reg-legged seriema (*Cariama cristata*), scarlet ibis (*Eudocimus ruber*), guira cuckoo (*Guira guira*), black-crowned night heron (*Nycticorax nycticorax*), southern wigeon (*Mareca sibilatrix*), ocellated turkey (*Meleagris ocellata*), roseate spoonbill (*Platalea ajaja*), glossy ibis (*Plegadis falcinellus*), and green aracari (*Pteroglossus viridis*)) and reptiles (green iguana (*Iguana iguana*) and red-eared slider (*Trachemys scrypta scrypta*)). This type of natural-looking outdoor habitat is a creative outlet for physical activity and mental stimulation, providing several environmental enrichment opportunities covering both generic (choice, control, variety, and complexity) and specific needs [21,24]. However, one of the logistical challenges of this type of enclosure involves the difficult access to the animals, so the group of owls in *Zoomarine* was trained to enter a crate, allowing easy access for medical reasons whenever needed. This played a key part, as these owls were under a preventative health program, which included but was not limited to fecal sampling for coprological examination (wet mount, Gram stain, Diff-Quik stain, and flotation). However, even under a comprehensive medical protocol, the untimely and low sensitivity of certain diagnostic tools may delay treatment and worsen the overall prognosis of a certain parasitic infection [25,26]. Considering this diagnosis challenge along with the characteristics of the mixed-species outdoor enclosure and the possibility of nearby access for wild fauna and pathogens cross-transmission, all birds were under a deworming protocol every six months, which consisted of fenbendazole (50 mg/kg, *per os* (PO), once), ivermectin (0.2 mg/kg, PO, once, 15 days after fenbendazole) and toltrazuril (10 mg/kg, PO, three administrations, every other day (EOD), 7 days after ivermectin).

Even considering the regular anti-parasitic protocol, respiratory parasitosis was included as a differential diagnosis after the development of respiratory signs in 45% of the owl population. Both the clinical signs and findings from the physical examination and complementary exams fall inside the non-specific signs of respiratory disease in birds [27,28]. Considering the complexity of the avian respiratory tract, disease processes can be located in different anatomic regions, though the distinction between clinical signs is not clear, as many of them (e.g., dyspnea, breathing with an extended neck, tachypnea, respiratory noises, and a change in pitch or voice) may refer to both upper and lower respiratory tract disease. However, upper airway disease does not usually present with severe respiratory distress [27,28]. Due to its specific anatomic characteristics (e.g., air sacs), the avian respiratory system constitutes a substantial target for infection by a myriad of infectious agents [27]. Previous authors reported unspecific clinical signs of infections by *Cyathostoma* sp., such as depression and sudden death, along with syndromes and lesions, such as pneumonia, bronchitis, and air sacculitis, in birds of prey [10,11,13,29].

Hematology abnormalities may include leukocytosis, whereas serum biochemistries are not generally of particular interest in respiratory/parasitic diseases [27]. Indeed, only two individuals, one of them with eosinophilia, showed a slight increase in their total leukocyte counts. Although eosinophilia is commonly associated with mammals with parasitic diseases, this relation in avian species is not straightforward, so more complementary studies are needed on this subject [30]. A physical examination confirmed one case of severe and generalized subcutaneous emphysema, with the typical crackling and air-filled distention, with the common causes being trauma or lung/air sac inflammation associated with parasites or other infectious agents [31–33].

The treatment protocol was adapted according to each individual's clinical condition. The description of the specific clinical approach falls beyond the scope of this work. It is important to note, however, the use of a broad-spectrum protocol that included fenbendazole (50 mg/kg, PO, once a day (SID), 3 to 5 days), in conjunction in some cases with ivermectin (0.2 mg/kg, PO, once, 5 to 10 days after the therapeutic protocol with fenbendazole). Metaphylaxis measures through the same deworming protocol may have prevented the development of a clinical parasitosis in the rest of the owl population. Only one of the five individuals that presented with clinical signs died after a hyperacute clinical respiratory presentation. The literature includes several reports of mortality in birds of prey infected with *Cyathostoma* sp. Fatal parasitic pneumonia was reported in three injured wild owls in Southern Ontario along with four juvenile *A. cunicularia* bred in captivity [10]. A survey in Canada of 394 specimens of Falconiformes and Strigiformes also described fatal infections [13]. Finally, a case of fatal epicarditis associated with *Cyathostoma* species in a hen harrier (*Circus cyaneus*) was also reported [14].

Up until now, the identification of *Cyathostoma* (*Hovorkonema*) nematodes collected from the respiratory tracts of Accipitriformes, Falconiformes, and Strigiformes in Europe and North America has been highly inconsistent among authors [8]. These inconsistencies arose due to several shifts in the taxonomic placement of *Cyathostoma* species, which stem from different interpretations of morphological characteristics, various revisions, and questionable synonymities [34–37]. Furthermore, these discrepancies have been compounded by different authors utilizing distinct classification systems. For instance, Hartwich amalgamated three C. (*Hovorkonema*) species—C. (*Hovorkonema*) *americana*, C. (*Hovorkonema*) *bronchialis*, and C. (*Hovorkonema*) *variegatum* sensu stricto—into C. (*Hovorkonema*) *variegatum* [37]. This conflation resulted in several identifications of C. (*Hovorkonema*) *variegatum* in birds of prey that corresponded to the C. (*Hovorkonema*) *americana* species [8].

C. (*Hovorkonema*) *americana*, which is a typical parasite of the trachea and air sacs of raptors, C. (*Hovorkonema*) *bronchialis*, typically found in Anseriformes and also detected in Casuariiformes, and C. (*Hovorkonema*) *variegatum*, a typical parasite of the trachea of cranes and storks, constitute the three species within this subgenus. According to Kanarek's comprehensive morphological and molecular examination of *Cyathostoma* nematodes parasitizing the respiratory tracts of birds of prey in Europe and North America, only one species of C. (*Hovorkonema*) in birds of prey is confirmed to exist: C. (*Hovorkonema*) *americana* [8]. However, the possible but rare incidental occurrence of other C. (*Hovorkonema*) species, typical for different avian host groups, is acknowledged.

With respect to our samples and focusing on the characteristics highlighted as the most reliable for *Cyathostoma* sp. identification, the subgenus C. (*Hovorkonema*) is defined by spicules measuring 0.45–0.8 mm, while *Cyathostoma* (*Cyathostoma*) (Blanchard, 1849) has measurements ranging from 0.08 to 0.4 mm [38]. Our specimens fall between these ranges (0.435 mm). However, the absence of thorn-like projections on the copulatory bursa aligns with classification under the C. (*Hovorkonema*) subgenus.

Two of the three species of this subgenera can be clearly distinguished based on spicular length. C. (*Hovorkonema*) *variegatum* boasts notably longer spicules (0.58–0.77 mm) [34,39], compared to C. (*Hovorkonema*) *americana* (0.47–0.511 mm) [16,40]. The spicules length of C. (*Hovorkonema*) *bronchialis* (0.51–0.7 mm) is closely aligned with that of C. (*Hovorkonema*) *variegatum* [16,41–43]. Although this matter is still in need of further research, it suggests the potential synonymy of C. (*Hovorkonema*) *bronchialis* with C. (*Hovorkonema*) *variegatum*, as theorized by Kanarek et al. [8].

Considering these observed spicule lengths and given the close similarity of C. (*Hovorkonema*) *americana*'s spicules to our exemplars, along with the fact that it is the parasite of our designated host type (birds of prey), the sole plausible identification is C. (*Hovorkonema*) *americana*. Furthermore, the total lengths of the male and female parasites also fall within the species' described measurements.

While C. (*Hovorkonema*) *variegatum* in typical hosts exclusively occurs in the trachea, C. (*Hovorkonema*) *americana* in birds of prey was mainly recorded in air sacs and less frequently in the trachea [8,10,11,34,39,40]. There are reports on the presence of *Cyathostoma* nematodes in the trachea, bronchi, and air sacs of birds of prey [12,36]. However, there is no specific mention of these parasites being found in the lung parenchyma, least of all exclusively in this site, like in the present case [8].

Hunter et al. reported the presence of a *Cyathostoma* species parasite in an *A. cunicularia* specimen; however, the species of the parasite was not specified [10]. To the best of our knowledge, the current report is the first description of *Cyathostoma* (*Hovorkonema*) *americana* in *A. cunicularia*. The morphological differences and lack of previous reports do not rule out, but rather suggest, that these parasites could be a new species, with C. (*Hovorkonema*) *americana* being the closest match among the described species. Further research is needed to confirm this hypothesis, including the molecular characterization of these parasites, comparing the results with the molecular analyses available in the literature [8,11].

Although there are several studies available on parasitism in wild animals, more specifically wild birds, the possibility of free-ranging avifauna acting as a source of infection for captive specimens is a rarely explored subject [25,44]. It has been established, however, that exposure to wild avifauna may result in the parasitism of captive birds, depending if certain conditions for parasitic development are suitable such as parasite–host specificity, parasite life cycle, host resistance, husbandry-related factors, and environmental factors [25,44]. The host specificity of parasites is variable. While some parasite species are only found in a limited number of host species (otherwise known as highly host-specific), other parasites can affect multiple host species (considered to be host generalists, with low host specificity) [44].

There was a potential origin of the parasitic infection in the *A. cunicularia* population, namely, a group of more than 50 wild cattle egrets (*B. ibis*) that used to gather on top of the enclosure net, attracted by individuals of the same species that were part of the zoological collection and in that same enclosure. This gathering followed years of a particularly high birth rate of the captive *B. ibis* population and an increase in the total number of individuals kept in the enclosure. The wild *B. ibis* built nests and stayed above the enclosure throughout the day, and several nestlings and fledglings were seen throughout this period. Moreover, an estimated count of these wild *B. ibis*, not immediately above the enclosure but in the adjacent trees, included more than 600 animals.

Cattle egrets are free-range migratory birds in the family Ardeidae (order Pelecaniformes) with widespread distribution. Inclusively, they are the only member of the genus *Bubulcus* found in the tropics, subtropics, and warm temperate zones [45,46]. They are frequently found in association with cattle, with whom they establish a symbiotic relationship. This species has an extremely large range and does not approach the thresholds for vulnerable conservation status under the range size criterion—cattle egrets are classified as Least Concern under the IUCN conservation status [45,46].

Even though metaphylaxis measures were adopted, and all cleaning and disinfection protocols for the owls' underground burrows were revised, the incidence of reinfection was high. The complete clinical resolution of the parasitic infection of the population of burrowing owls was only accomplished when all the resident cattle egrets were transferred to other zoological institutions, and their wild counterparts immediately stopped congregating above the enclosure. This probable focus of infection cannot, however, be confirmed since a coprological analysis of these wild individuals was not performed. Cattle egrets are not yet known to be definitive hosts of C. (*Hovorkonema*) *americana*, but several nematodes were reported in these migratory birds, including from the Syngamidae family [46]. In this study, 40 free-ranging specimens of *B. ibis* were captured and examined for endoparasites and associated ectoparasites. The gastrointestinal content of each specimen was examined for adults and eggs of different helminths and protozoan cysts. The observed endoparasites included *Ascaridia galli* (with the highest prevalence, 35%), *Heterakis* spp. (17.5%), *Capillaria* spp. (12.5%), *Trichostrongylus tenuis* (12.5%), *Fascioloides magna* (10%), and *Syngamus trachea* (7.5%). This study also showed that there was a significantly higher risk of co-infection by the different types of parasites and that the ubiquitous presence of *B. ibis* may play an important in parasite dispersal in a global perspective. Furthermore, cattle egrets feed daily on a variety of insects, crustaceans, fish, lizards, frogs, spiders, moths, rodents, and earthworms, making it a favorable host through the ingestion of intermediate and paratenic hosts [46,47].

Interestingly, Syngamidae nematodes, including *Cyathostoma* sp., have been reported in free-ranging birds from across Europe, yet, there is only one available occurrence of Syngamidae in wild avifauna in Portugal, namely, the genus *Syngamus*. This study was developed in a wildlife rehabilitation center in Castelo Branco, Portugal, from January 2016 to May 2017, concerning the examination of the fecal samples of a total of 65 birds, in which birds of prey were the most representative group (77% of the samples) [11,34,37,39,40,43,48].

Generalist parasites that tolerate host physiology differences can infect a large variety of avian species [44]. In the present case of monoxenous parasites, owls may have become directly infected by eating soil or other materials contaminated with the feces of wild avifauna. Moreover, since burrowing owls are territorial and tend to stay in the same area, in or close to their underground burrows, there might have been a higher exposure to the feces of the wild *B. ibis* population. The possibility of paratenic hosts (e.g., earthworms) facilitating the chance of infection cannot be excluded, since the owls' habitat included areas of earthy soil. Finally, reinfection is possible through the ingestion of food/substrates contaminated with their own feces.

Parasites seldom lead to subclinical effects in wild avifauna, thus contributing to the dispersion and transmission to birds under human care inhabiting enclosures that allow close proximity to wild birds or paratenic hosts [25,44]. This report highlights the need to consider external sources of excretion and parasitic contamination as part of the management of parasitic infections when the type of zoological enclosures allows the possibility of opportunistic parasitism in susceptible avian collections. More studies are needed demonstrating cases of parasite transmission in a zoological context and describing the species of parasites affecting both captive and free-ranging birds. More studies on zoo–wildlife cross-transmission would benefit the adaptation of specific environment control measures (e.g., architectural designing; the removal of nearby roosts; paratenic host–control programs), as well as enhance effective preventative medical and husbandry protocols.

Author Contributions: Conceptualization, G.N.M. and L.M.d.C.; methodology, G.N.M., J.T.C. and M.L.; validation, G.N.M., J.T.C. and M.P.; formal analysis, G.N.M. and J.T.C.; investigation, G.N.M., J.T.C., M.P. and M.L.; writing—original draft preparation, G.N.M., J.T.C., M.P. and C.F.; writing—review and editing, G.N.M., J.T.C., M.P., M.L., C.F., N.U. and L.M.d.C.; supervision, L.M.d.C.; project administration, G.N.M.; funding acquisition, J.T.C. and L.M.d.C. All authors have read and agreed to the published version of the manuscript.

Funding: The research developed by J.T.C., M.P., and L.M.d.C. is supported by national funds through FCT—Fundação para a Ciência e Tecnologia, under the scope of projects UIDB/00276/2020 (CIISA-FMV-ULisboa) and LA/P/0059/2020 (AL4AnimalS). J.T.C. holds a scientific initiation grant (ref. UIDB/00276/2020) and a master grant, MSC22Jul-04, both provided by CIISA-FMV-ULisboa.

Institutional Review Board Statement: Ethical review and approval were waived for this study, due to the absence of interference with the normal animal health program designed by the zoo's veterinarians. This study followed the daily activity of the institution and its normal animal management, in strict collaboration with the zoo owners and assistant veterinarians. No interferences were made during the regular health management of all collections, since all animals were dewormed, and fecal samples were mostly collected from soil after natural excretion by the animals.

Informed Consent Statement: Not applicable.

Data Availability Statement: The authors confirm that the data supporting the findings of this study are available within the article.

Acknowledgments: The authors would like to thank the veterinary nurses and all the bird trainers of *Zoomarine* for their fundamental work and dedication throughout this clinical situation.

Conflicts of Interest: The authors declare no conflict of interest. The funders had no role in the design of the study; in the collection, analyses, or interpretation of data; in the writing of the manuscript; or in the decision to publish the results.

References

1. The IUCN Red List of Threatened Species Home Page. BirdLife International 2016, *Athene cunicularia*. Available online: https://www.iucnredlist.org/species/pdf/93227732 (accessed on 6 July 2023).
2. Rosenberg, K.V.; Kennedy, J.A.; Dettmers, R.; Ford, R.P.; Reynolds, D.; Alexander, J.D.; Beardmore, C.J.; Blancher, P.J.; Bogart, R.E.; Butcher, G.S.; et al. Partners in Flight Landbird Conservation Plan: 2016 Revision for Canada and Continental United States; Bogart, R.E., Ed.; Partners in Flight. 2016, pp. 1–105. Available online: https://partnersinflight.org/resources/the-plan/ (accessed on 27 November 2023).
3. Brandes, S. Athene cunicularia, Animal Diversity Web. Available online: http://animaldiversity.org/accounts/Athene_cunicularia/ (accessed on 12 July 2023).
4. Haug, E.; Oliphant, L. Movements, activity patterns, and habitat use of Burrowing Owls in Saskatchewan. *J. Wildl. Manag.* **1990**, *54*, 27–35. [CrossRef]
5. Sarno, R.J.; Nixon, P.A.; Mealey, B.K.; Concoby, R.E.; Mrykalo, R.J.; Grigione, M.M. Suitability of translocation sites for Florida burrowing owls: Prey availability and diet. *Southeast. Nat.* **2012**, *11*, 755–764. [CrossRef]
6. Trulio, L.; Higgins, P. The diet of western burrowing owls in an urban landscape. *West. N. Am. Nat.* **2012**, *73*, 348–356. [CrossRef]
7. BBL—Longevity Records of North American Birds. Available online: http://genomics.senescence.info/species/biblio.php?id=427 (accessed on 23 October 2014).
8. Kanarek, G.; Zaleśny, G.; Sitko, J.; Blanco, A.I. Taxonomic status of *Cyathostoma* nematodes (Nematoda: Syngaminae) parasitizing respiratory tracts of birds of prey and owls in Europe and North America: How many species are there? *Helminthologia* **2016**, *53*, 47–54. [CrossRef]
9. Fernando, M.A.; Barta, J.R. Tracheal Worms. In *Parasitic Diseases of Wild Birds*, 1st ed.; Atkinson, C.T., Thomas, N.J., Hunter, D.B., Eds.; John Wiley & Sons, Ltd.: Ames, IA, USA, 2008; pp. 343–354, ISBN 978-0-813-82081-1.
10. Hunter, D.B.; McKeever, K.; Bartlett, C. Cyathostoma Infections in Screech Owls, Saw-whet Owls, and Burrowing Owls in Southern Ontario. In *Raptor Biomedicine*, 1st ed.; Redig, P.T., Cooper, J.E., Remple, J.D., Hunder, D.B., Eds.; University of Minnesota Press: Minneapolis, MN, USA, 1993; pp. 54–56, ISBN 978-0816622191.
11. Krone, O.; Friedrich, D.; Honisch, M. Specific status and pathogenicity of syngamid nematodes in bird species (Ciconiformes, Falconiformes, Gruiformes) from Germany. *J. Helminthol.* **2007**, *81*, 67–73. [CrossRef]
12. Kanarek, G.; Horne, E.C.; Zaleśny, G. *Cyathostoma (Cyathostoma) phenisci* Baudet, 1937 (Nematoda: Syngamidae), a parasite of respiratory tract of African penguin *Spheniscus demersus*: Morphological and molecular characterization with some ecological and veterinary notes. *Parasitol. Int.* **2013**, *62*, 416–422. [CrossRef] [PubMed]
13. Lavoie, M.; Mikaelian, I.; Sterner, M.; Villeneuve, A.; Fitzgerald, G.; McLaughlin, J.; Lair, S.; Martineau, D. Respiratory Nematodiases in Raptors in Quebec. *J. Wildl. Dis.* **1999**, *35*, 375–380. [CrossRef]
14. Vaughan-Higgins, R.; Murphy, S.; Carter, I.; Pocknell, A.; Harris, E.; Sainsbury, A.W. Fatal epicarditis in a hen harrier (*Circus cyaneus*) a red-listed bird of high conservation concern in Britain associated with *Cyathostoma* species and *Escherichia coli* infection. *Vet. Rec.* **2013**, *173*, 477. [CrossRef]
15. Lichenfels, J.R. Keys to the genera of the superfamily Strongyloidea. In *CIH Keys to the Nematode Parasites of Vertebrates;* Anderson, R.C., Chabaud, A.G., Willmott, S., Eds.; Commonwealth Institute of Helminthology: Farnham Royal, UK, 1980; Volume 7, 41p.
16. Chapin, E.A. Review of the nematode genera *Syngamus* Sieb. and *Cyathostoma* E. Blanch. *J. Agric. Res.* **1925**, *30*, 557–570.
17. Borgsteede, F.H.; Okulewicz, A.; Zoun, P.M.; Okulewicz, J. The helminth fauna of birds of prey (Accipitriformes, Falconiformes and Strigiformes) in the Netherlands. *Acta Parasitol.* **2003**, *48*, 200–207.
18. Stiko, J.; Okulewicz, A. *Checklist of the Nematodes in Birds in the Czech Republic and the Slovak Republic;* Comenius Museum: Přerov, Czech Republic, 2010; 104p, ISBN 9788090371392.
19. Kells, N.J. Review: The Five Domains model and promoting positive welfare in pigs. *Animals* **2022**, *16*, 100378. [CrossRef] [PubMed]
20. Mellor, D.J.; Beausoleil, N.J.; Littlewood, K.E.; McLean, A.N.; McGreevy, P.D.; Jones, B.; Wilkins, C. The 2020 Five Domains Model: Including Human–Animal Interactions in Assessments of Animal Welfare. *Animals* **2020**, *10*, 1870. [CrossRef] [PubMed]
21. Tallo-Parra, O.; Salas, M.; Manteca, X. Zoo Animal Welfare Assessment: Where Do We Stand? *Animals* **2023**, *13*, 1966. [CrossRef] [PubMed]
22. Panayotova-Pencheva, M.S. Parasites in Captive Animals: A Review of Studies in Some European Zoos. *Zool. Gart.* **2013**, *82*, 60–71. [CrossRef]
23. Papini, R.; Girivetto, M.; Marangi, M.; Mancianti, F.; Giangaspero, A. Endoparasite Infections in Pet and Zoo Birds in Italy. *Sci. World J.* **2012**, *2012*, 253127. [CrossRef] [PubMed]
24. Mellen, J.; MacPhee, M.S. Philosophy of Environmental Enrichment: Past, Present, and Future. *Zoo Biol.* **2001**, *20*, 211–226. [CrossRef]
25. Carrera-Játiva, P.D.; Morgan, E.R.; Barrows, M.; Wronski, T. Gastrointestinal parasites in captive and free-ranging birds and potential cross-transmission in a zoo environment. *J. Zoo Wildl. Med.* **2018**, *49*, 116–128. [CrossRef]

26. Carvalho, A.F.; Barnabe, A.S.; Federsoni, I.P.; Ferraz, R.R.; Marco, R.M.; Garcia, I.P. Efficacy of parasitological diagnosis methods in wild animals kept in captivity. *Arq. Do Inst. Biol.* **2017**, *84*, e0502016. [CrossRef]
27. Crosta, L. Respiratory Diseases of Parrots: Anatomy, Physiology, Diagnosis and Treatment. *Vet. Clin. Exot. Anim. Pract.* **2021**, *24*, 397–418. [CrossRef]
28. Forbes, N.A. Raptors: Parasitic disease. In *BSAVA Manual of Raptors, Pigeons and Passerine Birds*, 1st ed.; Chitty, J., Lierz, M., Eds.; British Small Animal Veterinary Association: Gloucester, UK, 2008; Chapter 19; pp. 205–206, ISBN 978-1-905319-04-6.
29. Lierz, M.; Schuster, R.; Ehrlein, J.; Gobel, T. Findings of *Hovorkonema variegatum* in northern goshawk (*Accipiter gentilis*). *Kleintierpraxis* **1998**, *43*, 43–46.
30. Mitchell, E.; Johns, J. Avian Hematology and Related Disorders. *Vet. Clin. Exot. Anim. Pract.* **2008**, *11*, 501–522. [CrossRef] [PubMed]
31. Bowles, H.L.; Odberg, E.; Harrison, G.J.; Kottwitz, J.J. Surgical resolution of soft tissue disorders. In *Clinical Avian Medicine*, 2nd ed.; Harrison, G.J., Lightfoot, T.L., Eds.; Spix Publishing, Inc.: Palm Beach, FL, USA, 2006; Volume 2, p. 794, ISBN 00-9754994-0-8.
32. Doneley, B.; Harrison, G.J.; Lightfoot, T.L. Maximizing Information from the Physical Examination. In *Clinical Avian Medicine*, 2nd ed.; Harrison, G.J., Lightfoot, T.L., Eds.; Spix Publishing, Inc.: Palm Beach, FL, USA, 2006; Volume 1, p. 169, ISBN 00-9754994-0-8.
33. Lee, S.-Y.; Kim, H.-J.; Kim, J.-W. Deflation Treatment for Subcutaneous Emphysema in a Goffin Cockatoo (*Cacatua goffini*). *J. Vet. Clin.* **2011**, *28*, 519–521.
34. Vogel, H. About biology and morphology of *Cyathostoma variegatum* (Creplin, 1849) [*Syngamus variegatum*]. *Z. Infekt.-Skrankh* **1928**, *34*, 97–117.
35. Ali, M.M. A review and revision of the subfamily Cyathostominae Nicoll, 1927 (Nematoda, Syngamidae). *Acta Parasitol. Pol.* **1970**, *17*, 237–246.
36. Ryzhikov, K.M. The biology of a nematode *Cyathostoma bronchialis* (Syngamidae: Strongylata). *Helminthologia* **1980**, *17*, 241–244.
37. Hartwich, G.; Nematoda, I.I. Strongylida: Strongyloidea und Ancylostomatoidea. In *The Wildlife of Germany*; Gustav Fisher Verlag: Jena, Germany; Frankfurt, Germany, 1994; p. 157.
38. Kanarek, G. The occurrence of *Cyathostoma (Cyathostoma) microspiculum* (Skrjabin, 1913) (Nematoda: Syngamidae) in the great cormorant [*Phalacrocorax carbo* (L., 1758)] in north-eastern Poland. *J. Helminthol.* **2009**, *83*, 391–398. [CrossRef] [PubMed]
39. Okulewicz, A. *Cyathostoma variegatum* (Creplin, 1849) Chapin, 1925 and *Cyathostoma lari* Blanchard, 1849 (Nematoda, Syngamidae) in Poland. *Wiad. Parazytol.* **1984**, *30*, 53–56.
40. Borgsteede, F.H.; Okulewicz, A. Justification of the species *Cyathostoma (Hovorkonema) americana*. *Helminthologia* **2001**, *38*, 115–121.
41. Griffiths, H.J.; Leary, R.M.; Fenstermacher, R. A new record for gapeworm *(Cyathostoma bronchialis)* infection of domestic geese in North America. *Am. J. Vet. Res.* **1954**, *15*, 298–299.
42. Ryžikov, K.M.; Zavadil, R. On the identity of species of the genus *Cyathostoma* collected from ostrich. *Acta Univ. Agric. Silvic. Brno* **1958**, *2*, 125–132.
43. Hernandez-Rodriguez, S.; Calerto-Carretero, R.; Becerra-Martel, C.; Dominguez de Tena, M.; Moreno-Montañez, T.; Martinez-Gomez, F.; Barasona-Mata, J. *Cyathostoma bronchialis* (Mühling, 1884) Chapin, 1925 (Nematoda: Syngamidae) in *Anser anser*, first record in Spain. *Rev. Iber. Parasitol.* **1975**, *35*, 367–371.
44. Carrera-Játiva, P.D.; Morgan, E.R.; Barrows, M.; Jiménez-Uzcátegui, G.; Roosevelt, J.; Tituaña, A. Free-ranging avifauna as a source of generalist parasites for captive birds in zoological settings: An overview of parasite records and potential for cross-transmission. *J. Adv. Vet. Anim. Res.* **2020**, *7*, 482–500. [CrossRef] [PubMed]
45. The IUCN Red List of Threatened Species Home Page. BirdLife International 2016, *Bubulcus ibis*. Available online: https://www.iucnredlist.org/fr/species/22697109/155477521 (accessed on 27 November 2023).
46. Omonona, A.O.; Ademola, I.O.; Ogunrotimi, M.O. Parasites of cattle egrets (*Bubulcus ibis*) and the associated haematological and biochemical changes. *J. Environ. Ext* **2014**, *12*, 31–40.
47. Adediran, O.A.; Jubril, A.J.; Uwalaka, E.C.; Jarikre, T.A. Parasites, haematology and morphological changes in tissues of a migratory *Bubulcus ibis*. *J. Zool.* **2015**, *4*, 26–31. [CrossRef]
48. Pires, H.; Matos, A.; Brida, T.; Lopes, F.; Alho, A.M.; Madeira de Carvalho, L.M. Parasitology survey in wild birds kept for rehabilitation in central eastern Portugal. In Proceedings of the10th International Symposium on Wild Fauna, Vila Real, Portugal, 21–23 September 2017; p. 92.

Journal of
Zoological and Botanical Gardens

MDPI

Article

Testing Mini-FLOTAC for the Monitorization of Gastrointestinal Parasitic Infections in Birds Kept at Four Iberian Zoological Institutions

João Lozano [1,2,*], Cándido Pombo [3], Rami Salmo [3], Cristiana Cazapal-Monteiro [3], María Sol Arias [3], Daniela Carvalho [4], Madalena Lordelo [4], Augusto Batista [5], Rui Bernardino [6], Laura Rinaldi [7], Manuela Oliveira [1,2,8], Adolfo Paz-Silva [3] and Luís Madeira de Carvalho [1,2]

1 CIISA—Centre for Interdisciplinary Research in Animal Health, Faculty of Veterinary Medicine, University of Lisbon, Avenida da Universidade Técnica, 1300-477 Lisbon, Portugal
2 Associate Laboratory for Animal and Veterinary Sciences (AL4AnimalS), 1300-477 Lisbon, Portugal
3 Control of Parasites Research Group (COPAR, GI-2120), Department of Animal Pathology, Faculty of Veterinary, University of Santiago de Compostela, 27142 Lugo, Spain
4 LEAF—Linking Landscape, Environment, Agriculture and Food—Research Center, Instituto Superior de Agronomia, University of Lisbon, Tapada da Ajuda, 1349-017 Lisbon, Portugal
5 Quinta Pedagógica dos Olivais, Câmara Municipal de Lisboa, Rua Cidade de Lobito, Olivais Sul, 1800-088 Lisbon, Portugal
6 Lisbon Zoo, Veterinary Hospital, Estrada de Benfica, 1549-004 Lisbon, Portugal
7 Unit of Parasitology and Parasitic Diseases, Department of Veterinary Medicine and Animal Production, University of Naples Federico II, Via Federico Delpino, 80137 Naples, Italy
8 cE3c—Centre for Ecology, Evolution and Environmental Changes & CHANGE—Global Change and Sustainability Institute, Faculdade de Ciências, University of Lisbon, 1749-016 Lisbon, Portugal
* Correspondence: jlozano@fmv.ulisboa.pt

Abstract: Birds kept in zoological institutions are highly exposed to gastrointestinal (GI) parasitism caused by coccidia and nematodes. The current research aimed to characterize the avian GI parasitic fauna in several zoological collections in Portugal and Spain. During the full year of 2022, a total of 120 fecal samples were collected from four zoological institutions: Lisbon Zoo, Olivais Pedagogical Farm, and Avian Biodiversity Center (Lisbon, Portugal), and Avifauna park (Lugo, Spain). Analysis was conducted in domestic bird species (autochthonous and exotic poultry breeds), and 18 different exotic bird species like Galliformes (peacock, pheasant), Anseriformes (duck), Psittaciformes (parrot, macaw, cockatiel, parakeet, cockatoo), Coraciiformes (motmot), Charadriiformes (avocet), Strigiformes (owl), Phoenicopteriformes (flamingo), Struthioniformes (ostrich), Rheiformes (rhea), and Casuariiformes (emu, cassowary). Feces were processed using Mini-FLOTAC (MF), to identify parasitic forms and quantify their shedding (oocysts or eggs per gram of feces). Moreover, 15 fecal samples from pheasants were also processed using the McMaster method (McM), to compare the parasite shedding and frequencies between techniques. MF implementation allowed identification of coccidia infections in all bird collections. Also, peacocks of the Lisbon Zoo tested positive for *Trichostrongylus tenuis* and *Strongyloides pavonis*, and the exotic birds from Avifauna park were also positive for several nematode species, with *Ascaridia* sp., *Capillaria* sp., *Strongyloides* sp., and *Syngamus trachea* eggs being detected in pheasants' feces. Moreover, the analysis of pheasants' feces with MF detected prevalences of 33% for coccidia oocysts, and 47% for *Capillaria* sp. and *Ascaridia* sp. eggs, while McM detected prevalences of 13%, 27%, and 40% for the respective parasite taxa, with no differences being observed between methods ($p = 0.39$, $p = 0.45$, and $p = 0.50$, respectively). This research provided more scientific support regarding the importance of using Mini-FLOTAC in routine parasitological diagnosis in birds kept at zoological institutions.

Keywords: zoological institutions; birds; gastrointestinal parasites; Mini-FLOTAC

1. Introduction

Zoological institutions have several important roles in society and for animal wildlife, particularly visitors' education, research, and ex situ or in situ animal conservation, being of extreme importance especially for Vulnerable, Endangered, or Critically Endangered animal species or breeds [1]. According to the most recent edition of IUCN's Red List of Threatened Species, 13% of bird species are classified as "Threatened" and 6% as "Near Threatened", at European level [2], which highlights the importance of their conservation.

Among avian species, chickens account for the highest number of breeds at the risk of extinction (29%) [3], with examples including the Portuguese chicken breeds "Preta Lusitânica" (Lusitanian Black chicken), "Pedrês Portuguesa" (Portuguese Pedrês chicken), "Branca" (White chicken), and "Amarela" (Yellow chicken), which are currently kept at small-scale farms and zoological institutions [4,5]. In addition, the proportion of avian species with unknown status of conservation is even higher (64%) than the observed for mammals (55%) [3].

Several biotic and abiotic factors challenge the conservation of avian genetic resources. Despite all noteworthy efforts developed by zoological institutions to mimic natural habitats and maximize animals' welfare, their contact with the same environment for long periods of time and irregularity or absence of antiparasitic drug treatments may lead to a high exposure to several pathogens, particularly gastrointestinal (GI) parasites like coccidia, capillarids, ascarids, and strongylids [6–10].

Gastrointestinal parasites like coccidia (e.g., *Eimeria* spp. and *Isospora* spp.), and nematodes such as *Capillaria* spp., ascarids and heterakids (e.g., *Ascaridia* spp. and *Heterakis* spp., respectively), *Trichostrongylus tenuis*, and *Strongyloides* spp., as well as respiratory nematodes such as syngamids (e.g., *Syngamus trachea*), whose eggs are swallowed from the tracheal mucosa into the GI tract and expelled with feces, constitute the most commonly diagnosed endoparasites in domestic and exotic birds. Previous studies described their occurrence in Galliformes, Struthioniformes, Rheiformes, Casuariiformes, Anseriformes, Psittaciformes, Columbiformes, Passeriformes, Falconiformes, Accipitriformes, and Strigiformes kept at several ornithological collections across the globe, particularly in Portugal [11–13], Spain [14], Italy [9], Serbia [8], UK [6], Egypt [15], Brazil [16–18], Iran [19], and China [20].

Their coprological diagnosis can be performed using quantitative techniques like the McMaster and Modified McMaster methods [21,22], qualitative techniques such as Willis Flotation, Natural Sedimentation, and Fecal Cultures [21], and quantitative–qualitative techniques like FLOTAC and Mini-FLOTAC [23,24]. All these methodologies allow the dynamics of oocyst or egg shedding in animal collections to be followed, and thus to optimize antiparasitic drug treatments [21,25]. The two former techniques have shown very interesting results in the identification and quantification of coccidia and helminth infections in several animal species, particularly horses, ruminants, and pets, with overall higher sensitivity and precision in comparison with the McMaster method [23,24,26,27]. Mini-FLOTAC in particular has been recently tested in birds [28–30], and successfully implemented at field level for the routine diagnosis of GI parasitic infections in poultry, peacocks, and ratites [10,31].

The current study aimed to (i) implement Mini-FLOTAC in the diagnosis of GI parasitic infections in four ornithological collections in Portugal and Spain and (ii) compare the diagnosis outcomes between McMaster and Mini-FLOTAC techniques.

2. Materials and Methods

2.1. Zoological Institutions and Sampling Procedures

Between January–December 2022, a total of 120 fresh fecal samples of clinically normal birds were collected at three Portuguese zoological institutions located in Lisbon, specifically the Lisbon Zoo (38°44′38.535″ N; 9°10′16.793″ W), Olivais Pedagogical Farm (38°45′47.186″ N; 9°6′42.666″ W), and Avian Biodiversity Centre (38°42′26.933″ N; 9°10′56.264″ W), as well as at one Spanish ornithological institution located in Lugo, the Avifauna park (43°3′14.554″ N; 7°41′22.903″ W).

At the Lisbon Zoo, a total of 24 fecal samples were collected from individual enclosures containing chickens (*Gallus gallus domesticus*; n = 10 samples), turkeys (*Meleagris gallopavo domesticus*; n = 2), peacocks (*Pavo cristatus*; n = 5), greater rheas (*Rhea americana*; n = 4), emus (*Dromaius novaehollandiae*, n = 2) and cassowaries (*Casuarius casuarius*; n = 1).

At the Olivais Pedagogical Farm (OPF), a zoological institution which harbors autochthonous farm animal breeds for conservation and educational purposes, a total of 26 fecal samples were collected from laying hens of the four Portuguese autochthonous breeds: "Branca" (n = 6), "Amarela" (n = 6), "Preta Lusitânica" (n = 6), and "Pedrês Portuguesa" (n = 8), with each breed being kept in separate enclosures.

At the Avian Biodiversity Centre (ABC), a newly established ornithological collection belonging to the Higher Institute of Agronomy, University of Lisbon, a total of 24 fecal samples were collected from each of the following enclosures containing autochthonous and exotic *G. gallus domesticus* breeds (two samples per enclosure): enclosure 1—"Amarela" breed; 2—"Preta Lusitânica" breed; 3—"Pedrês Portuguesa" breed; 4—"Pedrês Portuguesa" and "Amarela" breeds (mixed); 5—"Branca" and "Phoenix" breeds (mixed); 6 – "Araucana" and "Japanese Silkie" breeds (mixed); 7—"White Phoenix" breed; 8—"Light Brahma" breed; 9—common backyard chickens; 10—"Dark Brahma" breed; 11—domestic quails; 12—"White Leghorn" and "Ayam Cemani" breeds (mixed).

Finally, a total of 46 fecal samples were collected at Avifauna park from individual enclosures containing the following exotic and wild bird species: greater rhea (*Rhea americana*; n = 3 samples), curassow (*Crax fasciolata*; n = 3), flamingo (*Phoenicopterus* spp.; n = 3), ruddy shelduck (*Tadorna ferruginea*; n = 2), western capercaillie (*Tetrao urogallus*; n = 3), black swan (*Cygnus atratus*; n = 3), Monk parakeet (*Myiopsitta monachus*; n = 1), eagle-owl (*Bubo bubo*; n = 1), cockatiel (*Nymphicus hollandicus*; n = 2), macaw (*Ara* spp.; n = 2), motmot (*Momotus momota*; n = 1), superb parrot (*Polytelis swainsonii*; n = 1), cockatoo (*Cacatua alba*; n = 1), eastern rosella (*Platycercus eximius*; n = 1), avocet (*Recuvirostra avosetta*; n = 2), and pheasants (Phasianinae; n = 17) (Figure 1).

Figure 1. Some of the domestic and exotic bird species included in this study: (**A**) cassowary, (**B**) ostrich, (**C**) emu, (**D**) cockatoo, (**E**) ruddy shelducks, (**F**) flamingos, (**G**) pheasant, (**H**) "Araucana" breed, (**I**) "Brahma" breed, (**J**) "Ayam Cemani" breed, (**K**) "Japanese Silkie" breed, (**L**) "White Phoenix" breed, (**M**) "Pedrês Portuguesa" breed (Portuguese Pedrês chicken), (**N**) "Amarela" breed (Yellow chicken), (**O**) "Preta Lusitânica" breed (Lusitanian Black chicken), (**P**) "Branca" breed (White chicken); photo credits: João Lozano (**A–G**,**K**,**O**,**P**) and Madalena Lordelo (**H–J**,**L–N**).

Birds' fecal samples were collected from the environment, with no direct manipulation of the animals. Feces were collected using individual plastic bags, which were then labelled and immediately transported inside a cooling bag to two laboratories: (i) feces from the Portuguese zoological institutions were stored and processed at the Laboratory of Parasitology and Parasitic Diseases of the Faculty of Veterinary Medicine, University of Lisbon (Portugal); feces from Avifauna park were analyzed at the Laboratory of Zoonoses of the Faculty of Veterinary Medicine, University of Santiago de Compostela (Spain). In both cases, feces were stored at 4 °C for a maximum of one week.

2.2. Coprological Techniques

2.2.1. Phase 1—General Coprological Diagnosis Using Mini-FLOTAC

All birds' feces were processed using the Mini-FLOTAC (MF) protocol for exotic animals (1:20 dilution), following the procedures previously described for birds [10,24]. For each sample, 2 g of feces were mixed with 38 mL of saturated sucrose solution (specific gravity 1.2) using the Fill-FLOTAC device. Then, the fecal suspension was directly transferred to the previously assembled reading chamber, and all coccidia oocysts and helminth eggs were identified and counted using an analytic sensitivity of 10 oocysts or eggs per gram of feces (OPG or EPG, respectively). The identification of parasitic forms was based on the descriptions from McDougald and Fitz-Coy [32], Yazwinski and Tucker [33], and Zajac and Conboy [34]. Coccidia oocysts' incubation in potassium dichromate was not performed in the current research, and thus their identification at genus level was only performed in oocysts which were already sporulated when feces were collected, being based on the oocyst size and number of sporocysts and sporozoites inside [32].

2.2.2. Phase 2—Using Pheasants' Feces to Compare Mini-FLOTAC and McMaster Methods

A total of 15 fecal samples from pheasants of the Avifauna park, which were processed with the Mini-FLOTAC method in phase 1 of this study, were also analyzed using the McMaster method (McM) to compare parasite prevalences and shedding obtained with both techniques. For each sample, a total of 3 g of feces were weighed and placed into a plastic flask, to which 42 mL of water were added (dilution 1:15). Feces were vigorously mixed and then filtered through a mesh with 150 µm pore diameter. The filtered suspension was transferred to 12 mL tubes and centrifuged at 2000 rpm for 5 min. Then, the supernatant was discarded, and the sediment mixed with 6 mL of saturated sodium chloride solution (specific gravity 1.2). The final suspension was transferred to a McMaster chamber, to count parasitic forms using an analytic sensitivity of 50 OPG/EPG, in a light microscope [21].

In this phase of the study, the fecal dilutions chosen for the Mini-FLOTAC and McMaster methods were in accordance with the manufacturer instructions and published literature for both methods [21,24], allowing reproduction of their expected analytical performances.

Since feces from all domestic and exotic birds included in this study had a maximum of 2–5 g, only one repetition was performed for each sample. Thus, prior to their mixing with the flotation solutions, feces were gently homogenized inside the respective plastic bags, to promote a complete distribution of all parasitic forms. Moreover, the general low quantity of individual feces allowed for the use of only 15 fecal samples from pheasants to compare the analytical performances of the Mini-FLOTAC and McMaster methods in the study's second phase. Out of 17 pheasant feces, only two samples had less than three grams remaining after being processed with the Mini-FLOTAC method, and thus were not used in this study's phase.

2.3. Statistical Analysis

The software Microsoft® Excel® (v2405; Microsoft Corporation, Redmond, WA, USA, 2023) was used for storing all parasitological data recorded for each ornithological institution, as well as for table and chart editing. Also, the software IBM® SPSS® Statistics (v27 for Windows; IBM Corporation, Armonk, NY, USA) was used for descriptive statistics (mean and standard errors), and to perform the following analysis: OPG and EPG data recorded in all bird collections were first subjected to normality analysis, using the Shapiro–Wilk test (n < 50 samples in each collection), which allowed for the conclusion that the data were not normally distributed ($p < 0.001$). Thus, parasite shedding data and prevalences obtained with Mini-FLOTAC and McMaster methods were compared using the Wilcoxon and Chi-square tests, respectively, with a significance level of $p < 0.05$.

3. Results

The implementation of the Mini-FLOTAC method allowed identification of coccidia oocysts in avian feces from all ornithological collections: the poultry and peacocks from the Lisbon Zoo (156 ± 104 OPG and 60 ± 60 OPG, respectively); the Portuguese autochthonous chicken breeds kept at OPF—"Preta Lusitânica" (88 ± 83 OPG), "Branca" (48 ± 25 OPG), "Amarela" (43 ± 34 OPG), and "Pedrês Portuguesa" (13 ± 6 OPG); the poultry of the ABC collection (663 ± 196 OPG), with emphasis on enclosures containing backyards chickens (5230 ± 1710 OPG), "Pedrês Portuguesa" (700 ± 40 OPG), "Amarela" (665 ± 155 OPG) and "Preta Lusitânica" (630 ± 110 OPG), and to a lesser extent at Avifauna park (Spain), particularly in avocets (20 ± 20 OPG), pheasants (11 ± 6 OPG), and western capercaillies (7 ± 3 OPG) (Table 1).

Table 1. Coccidia and helminths' fecal shedding levels (mean ± standard error) identified in different poultry breeds at two ornithological collections, Olivais Pedagogical Farm and Avian Biodiversity Centre, using the Mini-FLOTAC method.

Bird Collections	Poultry Breeds	Coccidia and Helminth Shedding
Olivais Pedagogical Farm	"Preta Lusitânica"	*Eimeria* sp.: 88 ± 83 OPG
	"Pedrês Portuguesa"	*Eimeria* sp.: 13 ± 6 OPG
	"Branca"	*Eimeria* sp.: 48 ± 25 OPG
	"Amarela"	*Eimeria* sp.: 43 ± 34 OPG/*A. galli*: 22 ± 22 EPG
Avian Biodiversity Centre	1—"Amarela"	*Eimeria* sp.: 665 ± 155 OPG
	2—"Preta Lusitânica"	*Eimeria* sp.: 630 ± 110 OPG
	3—"Pedrês Portuguesa"	*Eimeria* sp.: 700 ± 40 OPG
	4—"Pedrês Portuguesa" + "Amarela"	*Eimeria* sp.: 5 ± 5 OPG
	5—"White Phoenix" + "Branca"	*Eimeria* sp.: 585 ± 295 OPG
	6—"Araucana" + "Japanese Silkie"	*Eimeria* sp.: 20 OPG *
	7—"White Phoenix"	*Eimeria* sp.: 50 OPG *
	8—"Light Brahma"	*Eimeria* sp.: 10 OPG */*A. galli*: 25 ± 5 EPG
	9—Backyard chickens	*Eimeria* sp.: 5230 ± 1710 OPG
	10—"Dark Brahma"	*Eimeria* sp.: 60 ± 30 OPG
	11—Quails	-
	12—"White Leghorn" + "Ayam Cemani"	*Eimeria* sp.: 5 ± 5 OPG

* Standard error values were equal to zero.

Moreover, nematode eggs were also detected in feces from all bird collections, especially at Avifauna park, where the highest shedding of *Capillaria* sp. eggs was detected, particularly in Monk parakeets (260 EPG) and western capercaillies (103 ± 74 EPG), as well as *Ascaridia* sp. (72 ± 33 EPG), and a single sample containing *Syngamus trachea* eggs (10 EPG) was recorded in pheasants. Also, *Strongyloides* sp. infections were mostly detected in curassows of Avifauna park (33 ± 18 EPG), and peacocks of the Lisbon Zoo tested positive for *T. tenuis* (8 ± 8 EPG) (Table 2; Figure 2).

Regarding the comparisons of results obtained with both coprological methods, it was found that the coccidia, *Capillaria* sp., and *Ascaridia* sp. overall shedding detected in 15 pheasants achieved 13 ± 7 OPG and 13 ± 10 OPG, 47 ± 31 EPG and 60 ± 46 EPG, and 71 ± 38 EPG and 137 ± 51 EPG, using the Mini-FLOTAC and McMaster methods, respectively, with no significant differences between techniques ($p = 0.89$, $p = 0.94$, and $p = 0.13$, respectively).

Table 2. Coccidia and helminths' fecal shedding levels (mean ± standard error) identified in different avian species kept at the Lisbon Zoo and Avifauna park, using the Mini-FLOTAC method.

Bird Collections	Bird Species	Coccidia and Helminth Shedding
Lisbon Zoo	Chicken	*Eimeria* sp.: 156 ± 104 OPG
	Turkey	-
	Peacock	*Eimeria* sp.: 60 ± 60 OPG/*T. tenuis*: 8 ± 8 EPG/*Strongyloides* sp.: 2 ± 2 EPG
	Greater Rhea	-
	Emu	-
	Cassowary	-
Avifauna park	Greater Rhea	-
	Curassow	*Strongyloides* sp.: 33 ± 18 EPG
	Flamingo	*Strongyloides* sp.: 7 ± 7 EPG
	Ruddy Shelduck	Ascarids: 20 ± 20 EPG
	Western Capercaillie	Coccidia: 7 ± 3 OPG [a]/*Capillaria* sp.: 103 ± 74 EPG
	Black Swan	-
	Monk Parakeet	*Capillaria* sp.: 260 EPG [b]
	Eagle-Owl	-
	Cockatiel	-
	Parrot	-
	Motmot	*Capillaria* sp.: 40 EPG [b]/Ascarids: 10 EPG [b]
	Superb Parrot	-
	Cockatoo	-
	Eastern Rosella	-
	Avocet	Coccidia: 20 ± 20 OPG [a]
	Pheasant	Coccidia: 11 ± 6 OPG [a]/*Capillaria* sp.: 44 ± 27 EPG/Ascarids: 72 ± 33 EPG/*Strongyloides* sp.: 1 ± 1 EPG/*Syngamus trachea*: 1 ± 1 EPG

[a] Unsporulated oocysts. [b] Standard error values were equal to zero.

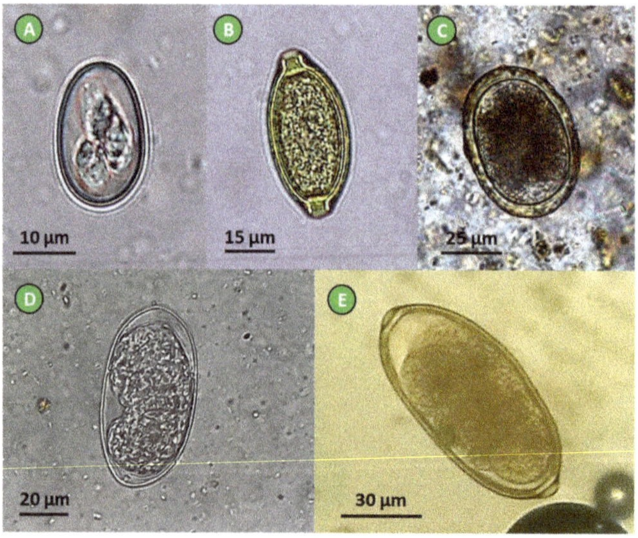

Figure 2. Parasites identified in feces of the selected avian species: (**A**) *Eimeria* sp. oocyst (chicken feces), (**B**) *Capillaria* sp. egg (pheasant feces), (**C**) *Ascaridia galli* egg (chicken feces), (**D**) *T. tenuis* egg (peacock feces), and (**E**) *Syngamus trachea* egg (pheasant feces); photos taken at 400× total magnification (credits: João Lozano).

The same scenario was observed for these parasites' prevalences, with the implementation of Mini-FLOTAC and McMaster methods allowing identification of prevalences of 33% and 13% for coccidia, 47% and 27% for *Capillaria* sp., and 47% and 40% for *Ascaridia* sp., respectively, with their differences being once again not significant ($p = 0.39$, $p = 0.45$, and $p = 0.50$, respectively) (Figure 3).

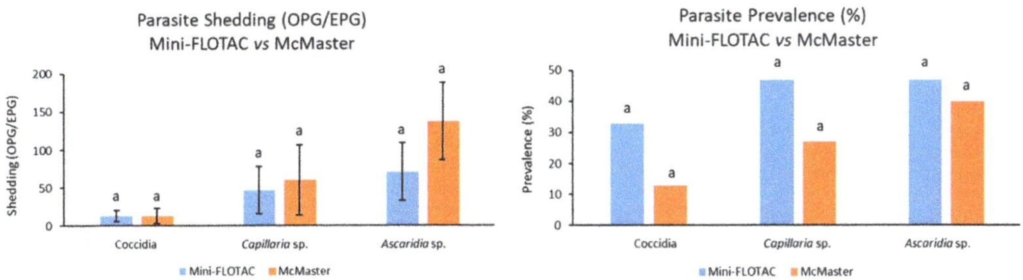

Figure 3. Comparison of coccidia, *Capillaria* sp., and *Ascaridia* sp. fecal shedding and prevalences between Mini-FLOTAC and McMaster methods; error bars on the left chart correspond to standard errors; equal letters on each bar mean no significant difference between techniques ($p > 0.05$).

4. Discussion

The control of animal GI parasitic infections still mainly relies in using anticoccidial and anthelminthic drugs, as well as live non-attenuated or live attenuated vaccines, with no previous laboratorial diagnosis [35–38]. However, this kind of approach is unsustainable, especially in zoological institutions, since antiparasitic drugs target mainly the endogenous stages of the parasites' life cycles and thus do not act on oocysts, eggs, and larvae accumulated in feces, soil, and/or pasture, and are also linked to drug resistance and bioaccumulation of pharmacological residues in soil and groundwaters [39–41]. Moreover, designing treatment protocols for exotic animal species is challenging for zoo veterinarians, since it often relies on adapting the guidelines reported for domestic animals, which can lead to low treatment efficacies, drug resistance, and even toxic effects in animals due to differences in the pharmacokinetics and pharmacodynamics of antiparasitic molecules between domestic and exotic animal species [42].

The implementation of the Mini-FLOTAC protocol for exotic animals allowed identification of different GI parasitic populations and respective shedding and prevalences in each bird collection. The identification of *Eimeria* sp. infections in laying hens belonging to Portuguese poultry breeds "Pedrês Portuguesa", "Preta Lusitânica", "Amarela", and "Branca", especially those kept at the Avian Biodiversity Centre, is in accordance with previous research performed in these poultry breeds kept free-ranging in a Portuguese organic farm, where authors also identified infections by *Eimeria* sp. and a respective oocyst fecal shedding up to 43,500 OPG [43]. Moreover, infections by this protozoan parasite were also identified in other Galliformes included in this study, particularly in birds of exotic breeds at the Avian Biodiversity Centre, as well as in chickens and peacocks of the Lisbon Zoo, and in pheasants of the Avifauna park, highlighting the susceptibility of birds from this order to become infected by sporulated coccidian oocysts, as reported in birds from other zoological institutions [6,8,11,12,14,15,20].

Coccidia are one of the most ubiquitous GI parasites in birds, with the genera *Eimeria* and *Isospora* having the most significant health and economic impacts on ornithological collections [32,37,44,45]. The combined effect of the intrinsic biological characteristics of this protozoan parasite, particularly its monoxenous and very short life cycle, with a pre-patent period of just 3–4 days [32,37,46], and the fact that zoo birds are often exposed to same outdoor environment for long periods, and consequently to sporulated oocysts in feces, soil, feed, and drinking water [32], leads to a continuous re-infection process. Although birds from the current study did not exhibit any clinical sign of coccidiosis, the detection of oocysts in their feces justifies the need for their regular monitoring, by means of coprological diagnosis, fecal quality, feather appearance analysis, and bird behavior observation. Clinical coccidiosis can include moderate to severe diarrhea episodes, sometimes with blood coagula, and may eventually lead to death if birds are not properly treated. Subclinical disease is often difficult to diagnose, being more linked to decreased nutrient absorption and consequently weight loss, anorexia, and prostration [37,46]. Thus, its control is of most importance in zoological institutions and conservation centers, especially in those harboring native and exotic poultry breeds, whose majority is near extinction [3]. The integration of anticoccidial drugs (coccidiostatic or coccidicidal drugs) with other control solutions, particularly vaccination, improved sanitizing, and the implementation of biological solutions such as feeding birds with prebiotics and probiotics, plant extracts, essential oils, and parasiticide fungi, have been proposed for the sustainable control of poultry coccidiosis [47–49].

Nematodes were also detected in all bird collections, with *A. galli* eggs being identified in feces from laying hens of the "Amarela" breed, a result already described in previous research in autochthonous poultry breeds [43], and in peacocks as previously reported at Brazilian [18] and Iranian zoos [19]. This ascarid is the largest roundworm of Galliformes, and its pathogenicity is associated to the thickening the intestinal epithelium and appearance of hemorrhagic spots along with edema, as well as ulcerative proventriculitis, leading to diarrhea and anemia, and with the accumulation of adult forms being capable of blocking the GI lumen in more severe infections [33,50–52], and thus being of clinical importance for ornithological collections.

Moreover, *Capillaria* sp. infections were recorded in exotic bird species of the Avifauna park, and despite the overall low shedding of this nematode in the majority of the host species, its identification in the western capercaillie should be underlined. Infections by this nematode in birds may cause catarrhal inflammation and thickening of the esophagus and crop epithelia, as well as hemorrhagic enteritis in small and large intestines, and anemia in more acute cases [33,53]. Since the population trend of the Western Capercaillie is currently classified as "Decreasing", according to the latest edition of the IUCN's Red List [2] and data from the European Nature Information System [54], the presence of this parasite in this bird species may pose future health and conservation concerns, with the regular monitoring of this nematode's egg excretion and detection of any clinical signs being of utmost importance for its control.

Also, pheasants of the Avifauna park tested positive for *S. trachea*, a nematode of the respiratory tract of birds and often responsible for serious health and economic concerns in wild and game pheasants, since nematode adult forms attach to the tracheal mucosa and, together with mucous production, lead to lumen obstruction and consequently to birds asphyxiation [33,55,56]. Thus, despite its low egg shedding, the identification of this nematode on these avian hosts justifies a higher awareness and their regular monitoring.

Results from the comparisons between Mini-FLOTAC and McMaster performances regarding parasite shedding or prevalence allowed determination of similar results for both techniques. Moreover, and despite the absence of statistical difference between the outputs obtained with each method, it must be noted that Mini-FLOTAC detected a coccidia prevalence 2.5× higher (33%) than the one obtained with the McMaster method (13%). Overall results from these comparisons allow the conclusion that Mini-FLOTAC is indeed a valuable alternative to the McMaster method for the diagnosis of coccidia and helminth infections in domestic and exotic birds, as reported in previous studies [10,11,24,31]. In addition, the fact that this method only comprises three devices for fecal processing and analysis—Fill-FLOTAC, reading chamber, and a light microscope—together with an extensively proven high sensitivity and precision [24,26] supports its use by zoo veterinarians for a simple, quick, non-expensive, and "in-house" parasitological diagnosis, as previously described in several zoological institutions [10,57,58]. Further research could also include the comparison of the McMaster method with other Mini-FLOTAC and FLOTAC dilutions [23,24], to optimize the coprological diagnosis of birds' GI parasitism.

To the authors best knowledge, research on the GI parasitism of autochthonous poultry breeds and exotic bird species is still scarce worldwide, and overall results from this study suggest the need for further broader-scale research in more public and private zoological collections, especially those harboring avian species facing the threat of extinction, and at a multi-country level, since the integration of regular monitoring of avian parasitism with antiparasitic drug treatments is of most importance for their conservation.

Author Contributions: Conceptualization, J.L., M.O., A.P.-S. and L.M.d.C.; methodology, J.L., C.P., R.S., C.C.-M., M.S.A., L.R., M.O., A.P.-S. and L.M.d.C.; software, J.L.; validation, M.O., A.P.-S. and L.M.d.C.; formal analysis, J.L.; investigation, J.L., M.O., A.P.-S. and L.M.d.C.; resources, J.L., C.P., R.S., C.C.-M., M.S.A., D.C., M.L., A.B., R.B., L.R., M.O., A.P.-S. and L.M.d.C.; data curation, J.L. and C.P.; writing—original draft preparation, J.L.; writing—review and editing, M.O., A.P.-S. and L.M.d.C.; visualization, J.L.; supervision, M.S.A., A.P.-S., M.L., A.B., R.B., L.R., M.O., A.P.-S. and L.M.d.C.; project administration, L.M.d.C.; funding acquisition, J.L., M.S.A., M.O., A.P.-S. and L.M.d.C. All authors have read and agreed to the published version of the manuscript.

Funding: This research was funded by CIISA/FMV Project UIDB/00276/2020 and LA/P/0059/2020—AL4AnimalS (both funded by FCT), as well as by Project ED431B 2021/07 (Consellería de Cultura, Educación e Universidades, Xunta de Galicia). Also, João Lozano holds a Ph.D. Research Fellowship, 2020.09037.BD (funded by FCT).

Institutional Review Board Statement: Ethical review and approval were waived for this study since fecal samples were collected from soil, after excretion, and thus without any direct manipulation of the birds. Moreover, all procedures were integrated in the normal daily activities of each zoological institution.

Data Availability Statement: The data presented in this study are available on request from the corresponding author.

Acknowledgments: The authors would like to acknowledge the owners of each bird collection for allowing us to perform fecal samplings at the respective facilities, and to thank all veterinarians and technicians for all the support provided during this research.

Conflicts of Interest: The authors declare no conflicts of interest.

References

1. Miranda, R.; Escribano, N.; Casas, M.; Pino-Del-Carpio, A.; Villarroya, A. The Role of Zoos and Aquariums in a Changing World. *Annu. Rev. Anim. Biosci.* **2023**, *11*, 287–306. [CrossRef] [PubMed]
2. IUCN Red List. European Red List of Birds 2021. Available online: https://data.europa.eu/doi/10.2779/959320 (accessed on 7 December 2023).
3. FAO. Status and Trends of Animal Genetic Resources. Available online: https://openknowledge.fao.org/items/1875926b-3ca0-4e9b-9012-465d7d6db9f2 (accessed on 15 May 2024).
4. Carolino, I.; Lopes, S.; Silva, F.; Oliveira e Sousa, C.; Almeida, J.; Carolino, N. Galinhas de Raças Autóctones e a Sua Importância Na Agricultura Sustentável. *Rev. Voz Do Campo* **2014**, 22–23. Available online: https://vozdocampo.pt/2022/05/15/galinhas-de-rac%CC%A7as-autoctones-e-a-sua-importa%CC%82ncia-na-agricultura-sustentavel/ (accessed on 4 June 2024).
5. Brito, N.; Lopes, J.; Ribeiro, V.; Dantas, R.; Leite, J. Small Scale Egg Production: The Challenge of Portuguese Autochthonous Chicken Breeds. *Agriculture* **2021**, *11*, 818. [CrossRef]
6. Carrera-Játiva, P.D.; Morgan, E.R.; Barrows, M.; Wronski, T. Gastrointestinal Parasites in Captive and Free-Ranging Birds and Potential Cross-Transmission in a Zoo Environment. *J. Zoo Wildl. Med.* **2018**, *49*, 116–128. [CrossRef] [PubMed]
7. Carrera-Játiva, P.D.; Morgan, E.R.; Barrows, M.; Jiménez-Uzcátegui, G.; Armijos Tituaña, J.R. Free-Ranging Avifauna as a Source of Generalist Parasites for Captive Birds in Zoological Settings: An Overview of Parasite Records and Potential for Cross-Transmission. *J. Adv. Vet. Anim. Res.* **2020**, *7*, 482–500. [CrossRef] [PubMed]
8. Ilić, T.; Becskei, Z.; Gajić, B.; Özvegy, J.; Stepanović, P.; Nenadović, K.; Dimitrijević, S. Prevalence of Endoparasitic Infections of Birds in Zoo Gardens in Serbia. *Acta Parasitol.* **2018**, *63*, 134–146. [CrossRef]
9. Papini, R.; Girivetto, M.; Marangi, M.; Mancianti, F.; Giangaspero, A. Endoparasite Infections in Pet and Zoo Birds in Italy. *Sci. World J.* **2012**, *2012*, 253127. [CrossRef] [PubMed]
10. Lozano, J.; Almeida, C.; Victório, A.; Melo, P.; Rodrigues, J.; Rinaldi, L.; Cringoli, G.; Gomes, L.; Oliveira, M.; Paz-Silva, A.; et al. Implementation of Mini-Flotac in Routine Diagnosis of Coccidia and Helminth Infections in Domestic and Exotic Birds. *Vet. Sci.* **2021**, *8*, 160. [CrossRef]

11. Rosa de Almeida, C. Caraterização Do Parasitismo Gastrointestinal Em Coleções de Pavão Comum (*Pavo cristatus*) Inseridos No Património Cultural Na Região de Lisboa. MSc Thesis, Faculty of Veterinary Medicine, University of Lisbon, Lisbon, Portugal, 2022.
12. Rocha de Castro, D. Pesquisa de Parasitas Gastrointestinais Em Aves de Rapina Mantidas Em Cativeiro Em Portugal. MSc Thesis, Faculty of Veterinary Medicine, University of Lisbon, Lisbon, Portugal, 2021.
13. Lozano, J.; Palomero, A.; Anaya, A.; Lux-Hoppe, E.; Gomes, L.; Oliveira, M.; Paz-Silva, A.; Teresa Rebelo, M.; Madeira de Carvalho, L. Long-Term Survey of *Eimeria* spp. Prevalence and Faecal Shedding in a Traditional Portuguese Free-Range Broiler Farm. *Rev. Port. Zoot* **2021**, *6*, 13–23.
14. Pérez Cordón, G.; Hitos Prados, A.; Romero, D.; Sánchez Moreno, M.; Pontes, A.; Osuna, A.; Rosales, M.J. Intestinal Parasitism in the Animals of the Zoological Garden "Peña Escrita" (Almuñecar, Spain). *Vet. Parasitol.* **2008**, *156*, 302–309. [CrossRef]
15. Kamel, A.; Abdel-Latef, G. Prevalence of Intestinal Parasites with Molecular Detection and Identification of *Giardia duodenalis* in Fecal Samples of Mammals, Birds and Zookeepers at Beni-Suef Zoo, Egypt. *J. Parasit. Dis.* **2021**, *45*, 695–705. [CrossRef]
16. Hofstatter, P.G.; Guaraldo, A.M.A. Levantamento Parasitológico Em Aves de Alguns Zoológicos Brasileiros. *Rev. Bras. De Parasitol. Vet.* **2015**, *24*, 87–91. [CrossRef]
17. Gallo, S.S.M.; Teixeira, C.S.; Ederli, N.B.; Oliveira, F.C.R. Gastrointestinal Parasites of a Population of Emus (*Dromaius novaehollandiae*) in Brazil. *Braz. J. Biol.* **2020**, *80*, 66–72. [CrossRef]
18. Melo, Y.; Ferraz, H.; Saturnino, K.; Silva, T.; Braga, I.; Amaral, A.; Meirelles-Bartoli, R.; Ramos, D. Gastrointestinal Parasites in Captive and Free-Living Wild Birds in Goiania Zoo. *Braz. J. Biol.* **2022**, *82*, e240386. [CrossRef] [PubMed]
19. Nasiri, V.; Jameie, F. Intestinal Parasitic Infection in Wild Animals of a Zoological Garden in Alborz, Iran. *J. Istanb. Vet. Sci.* **2019**, *3*, 37–42. [CrossRef]
20. Zhang, K.; Fu, Y.; Ma, C.; Wang, Y.; Zhang, J.; Lang, J.; Han, K.; Chen, Y.; Yu, F.; Zhang, L. First Detection of Coccidia Species in Blue Peafowl (*Pavo cristatus*) in China via Both Microscopic and Molecular Methods. *Vet. Parasitol. Reg. Stud. Rep.* **2022**, *36*, 100807. [CrossRef] [PubMed]
21. Zajac, A.; Conboy, G. Fecal Exam Procedures. In *Veterinary Clinical Parasitology*; Zajac, A., Conboy, G., Eds.; Wiley-Blackwell: Ames, Iowa, 2012; pp. 140–155.
22. Shapiro, L. Appendix I: Important Techniques for Veterinary Technicians. In *Pathology and Parasitology for Veterinary Technicians*; Shapiro, L., Ed.; Delmar, Cengage Learning: Boston, MA, USA, 2010; pp. 223–241.
23. Cringoli, G.; Rinaldi, L.; Maurelli, M.P.; Utzinger, J. FLOTAC: New Multivalent Techniques for Qualitative and Quantitative Copromicroscopic Diagnosis of Parasites in Animals and Humans. *Nat. Protoc.* **2010**, *5*, 503–515. [CrossRef] [PubMed]
24. Cringoli, G.; Maurelli, M.P.; Levecke, B.; Bosco, A.; Vercruysse, J.; Utzinger, J.; Rinaldi, L. The Mini-FLOTAC Technique for the Diagnosis of Helminth and Protozoan Infections in Humans and Animals. *Nat. Protoc.* **2017**, *12*, 1723–1732. [CrossRef] [PubMed]
25. Englar, R. Quick Assessment Tests (QATS) Involving Feces. In *Low-Cost Veterinary Clinical Diagnostics*; Englar, R., Dial, S., Eds.; John Wiley & Sons, Inc.: Hoboken, NJ, USA, 2023; pp. 159–233.
26. Maurelli, M.P.; Dourado Martins, O.M.; Morgan, E.R.; Charlier, J.; Cringoli, G.; Mateus, T.L.; Bacescu, B.; Chartier, C.; Claerebout, E.; de Waal, T.; et al. A Qualitative Market Analysis Applied to Mini-FLOTAC and Fill-FLOTAC for Diagnosis of Helminth Infections in Ruminants. *Front. Vet. Sci.* **2020**, *7*, 580649. [CrossRef]
27. Maurelli, M.P.; Rinaldi, L.; Alfano, S.; Pepe, P.; Coles, G.C.; Cringoli, G. Mini-FLOTAC, a New Tool for Copromicroscopic Diagnosis of Common Intestinal Nematodes in Dogs. *Parasit. Vectors* **2014**, *7*, 356. [CrossRef]
28. Coker, S.M.; Pomroy, W.E.; Howe, L.; McInnes, K.; Vallee, E.; Morgan, K.J. Comparing the Mini-FLOTAC and Centrifugal Faecal Flotation for the Detection of Coccidia (*Eimeria* spp.) in Kiwi (*Apteryx mantelli*). *Parasitol. Res.* **2020**, *119*, 4287–4290. [CrossRef]
29. Daş, G.; Klauser, S.; Stehr, M.; Tuchscherer, A.; Metges, C.C. Accuracy and Precision of McMaster and Mini-FLOTAC Egg Counting Techniques Using Egg-Spiked Faeces of Chickens and Two Different Flotation Fluids. *Vet. Parasitol.* **2020**, *283*, 109158. [CrossRef]
30. Bortoluzzi, C.; Paras, K.L.; Applegate, T.J.; Verocai, G.G. Comparison between McMaster and Mini-FLOTAC Methods for the Enumeration of Eimeria Maxima Oocysts in Poultry Excreta. *Vet. Parasitol.* **2018**, *254*, 21–25. [CrossRef]
31. Lozano, J.; Anaya, A.; Rinaldi, L.; Cringoli, G.; Gomes, L.; Oliveira, M.; Paz-Silva, A.; Rebelo, M.; Madeira de Carvalho, L. Diagnosis of Coccidiosis by *Eimeria* Spp. in Free-Range Chickens Using Mini-FLOTAC and McMaster Techniques-Preliminary Results. *Sci. Parasitol.* **2021**, *22*, 13–18.
32. McDougald, L.; Fitz-Coy, S. Parasitic Diseases—Coccidiosis. In *Diseases of Poultry*; Saif, Y., Ed.; Blackwell Publishing Professional: Ames, IA, USA, 2008; pp. 1068–1085.
33. Yazwinski, T.; Tucker, C. Nematodes and Acanthocephalans. In *Diseases of Poultry*; Saif, Y., Ed.; Blackwell Publishing Professional: Ames, IA, USA, 2008; pp. 1025–1066.
34. Zajac, A.; Conboy, G. Parasites of Domestic Animals—Birds. In *Veterinary Clinical Parasitology*; Zajac, A., Conboy, G., Eds.; Wiley-Blackwell: Ames, IA, USA, 2012; pp. 140–155.
35. Erez, M.S.; Doğan, İ.; Kozan, E.; Göksu, A. A Survey of Knowledge, Approaches, and Practices Surrounding Parasitic Infections and Antiparasitic Drug Usage by Veterinarians in Türkiye. *Animals* **2023**, *13*, 2693. [CrossRef]
36. Peek, H.W.; Landman, W.J.M. Coccidiosis in Poultry: Anticoccidial Products, Vaccines and Other Prevention Strategies. *Vet. Q.* **2011**, *31*, 143–161. [CrossRef] [PubMed]

37. Mesa-Pineda, C.; Navarro-Ruíz, J.L.; López-Osorio, S.; Chaparro-Gutiérrez, J.J.; Gómez-Osorio, L.M. Chicken Coccidiosis: From the Parasite Lifecycle to Control of the Disease. *Front. Vet. Sci.* **2021**, *8*, 787653. [CrossRef] [PubMed]
38. Zaheer, T.; Abbas, R.Z.; Imran, M.; Abbas, A.; Butt, A.; Aslam, S.; Ahmad, J. Vaccines against Chicken Coccidiosis with Particular Reference to Previous Decade: Progress, Challenges, and Opportunities. *Parasitol. Res.* **2022**, *121*, 2749–2763. [CrossRef] [PubMed]
39. Abbas, R.Z.; Iqbal, Z.; Blake, D.; Khan, M.N.; Saleemi, M.K. Anticoccidial Drug Resistance in Fowl Coccidia: The State of Play Revisited. *Worlds Poult. Sci. J.* **2011**, *67*, 337–349. [CrossRef]
40. Mooney, D.; Richards, K.G.; Danaher, M.; Grant, J.; Gill, L.; Mellander, P.E.; Coxon, C.E. An Analysis of the Spatio-Temporal Occurrence of Anthelmintic Veterinary Drug Residues in Groundwater. *Sci. Total Environ.* **2021**, *769*, 144804. [CrossRef]
41. Martins, R.R.; Silva, L.J.G.; Pereira, A.M.P.T.; Esteves, A.; Duarte, S.C.; Pena, A. Coccidiostats and Poultry: A Comprehensive Review and Current Legislation. *Foods* **2022**, *11*, 2738. [CrossRef] [PubMed]
42. Panayotova-Pencheva, M.S. Experience in the Ivermectin Treatment of Internal Parasites in Zoo and Captive Wild Animals: A Review. *Zool. Gart.* **2016**, *85*, 280–308. [CrossRef]
43. Carvalho, I.; Castanheira, F.; Mateus, T. Parasitas Gastrointestinais Em Aves de Produção Biológica e de Raça Autóctone Portuguesa. *Agrotec. Cuid. Veterinários* **2017**, *23*, 22–25.
44. Blake, D.P.; Knox, J.; Dehaeck, B.; Huntington, B.; Rathinam, T.; Ravipati, V.; Ayoade, S.; Gilbert, W.; Adebambo, A.O.; Jatau, I.D.; et al. Re-Calculating the Cost of Coccidiosis in Chickens. *Vet. Res.* **2020**, *51*, 115. [CrossRef]
45. Panayotova-Pencheva, M.S. Parasites in Captive Animals: A Review of Studies in Some European Zoos. *Zool. Gart.* **2013**, *82*, 60–71. [CrossRef]
46. Lozano, J.; Anaya, A.; Palomero Salinero, A.; Lux-Hoppe, E.; Gomes, L.; Paz-Silva, A.; Teresa Rebelo, M.; Madeira de Carvalho, L. Gastrointestinal Parasites of Free-Range Chickens—A Worldwide Issue. *Bull. Univ. Agric. Sci. Vet. Med. Cluj-Napoca. Vet. Med.* **2019**, *76*, 110. [CrossRef]
47. Attree, E.; Sanchez-Arsuaga, G.; Jones, M.; Xia, D.; Marugan-Hernandez, V.; Blake, D.; Tomley, F. Controlling the Causative Agents of Coccidiosis in Domestic Chickens; an Eye on the Past and Considerations for the Future. *CABI Agric. Biosci.* **2021**, *2*, 37. [CrossRef]
48. Jamil, M.; Aleem, M.T.; Shaukat, A.; Khan, A.; Mohsin, M.; Rehman, T.U.; Abbas, R.Z.; Saleemi, M.K.; Khatoon, A.; Babar, W.; et al. Medicinal Plants as an Alternative to Control Poultry Parasitic Diseases. *Life* **2022**, *12*, 449. [CrossRef] [PubMed]
49. Lozano, J.; Louro, M.; Almeida, C.; Victório, A.C.; Melo, P.; Rodrigues, J.P.; Oliveira, M.; Paz-Silva, A.; Madeira de Carvalho, L. Isolation of Saprophytic Filamentous Fungi from Avian Fecal Samples and Assessment of Its Predatory Activity on Coccidian Oocysts. *Sci. Rep.* **2023**, *13*, 8965. [CrossRef] [PubMed]
50. Brar, R.S.; Kumar, R.; Leishangthem, G.D.; Banga, H.S.; Singh, N.D.; Singh, H. *Ascaridia galli* Induced Ulcerative Proventriculitis in a Poultry Bird. *J. Parasit. Dis.* **2016**, *40*, 562–564. [CrossRef] [PubMed]
51. Shohana, N.N.; Rony, S.A.; Ali, M.H.; Hossain, M.S.; Labony, S.S.; Dey, A.R.; Farjana, T.; Alam, M.Z.; Alim, M.A.; Anisuzzaman. Ascaridia galli Infection in Chicken: Pathobiology and Immunological Orchestra. *Immun. Inflamm. Dis.* **2023**, *11*, e1001. [CrossRef] [PubMed]
52. Höglund, J.; Daş, G.; Tarbiat, B.; Geldhof, P.; Jansson, D.S.; Gauly, M. *Ascaridia galli*—An Old Problem That Requires New Solutions. *Int. J. Parasitol. Drugs Drug Resist.* **2023**, *23*, 1–9. [CrossRef] [PubMed]
53. Groves, P.J. Impact of Parasites on Australian Laying Hen Welfare. *Anim. Prod. Sci.* **2021**, *61*, 1031–1036. [CrossRef]
54. European Nature Information System. Capercaillie—Tetrao Urogallus (Linnaeus, 1758). Available online: https://eunis.eea.europa.eu/species/1315 (accessed on 26 January 2024).
55. Martin, K.A.; Jesudoss Chelladurai, J.R.J.; Sato, Y.; Brewer, M.T. *Syngamus* Asphyxiation in a Captive Ring-Necked Pheasant. *Vet. Parasitol. Reg. Stud. Rep.* **2020**, *22*, 100493. [CrossRef] [PubMed]
56. Friend, M.; Franson, J. *Field Manual of Wildlife Diseases—General Field Procedures and Diseases of Birds*, 1st ed.; USGS: Madison, WI, USA, 1999; ISBN 0607880961.
57. Capasso, M.; Maurelli, M.P.; Ianniello, D.; Alves, L.C.; Amadesi, A.; Laricchiuta, P.; Silvestre, P.; Campolo, M.; Cringoli, G.; Rinaldi, L. Use of Mini-FLOTAC and Fill-FLOTAC for Rapidly Diagnosing Parasitic Infections in Zoo Mammals. *Rev. Bras. De Parasitol. Vet.* **2019**, *28*, 168–171. [CrossRef] [PubMed]
58. Lourenço, M. Survey of Gastrointestinal Parasites of Non-Human Primates from Two Iberian Zoos and Perspectives of Their Integrated Control. MSc Thesis, Faculty of Veterinary Medicine, University of Lisbon, Lisbon, Portugal, 2023.

Journal of
Zoological and
Botanical Gardens

MDPI

Article

Coprological Survey of Helminths in Reindeer (*Rangifer tarandus*) in 50 Selected Zoos and Menageries in Russia

Olga A. Loginova [1,*], Svetlana V. Akulova [2], Dmitry N. Egorov [2], Natalia S. Egorova [2], Sergei E. Spiridonov [1], Iuliia K. Plotnikova [1], Larisa M. Belova [3], Yuriy E. Kuznetsov [3], Daria I. Chuprak [3], Anna A. Krutikova [3], Iuliia V. Vasilkova [4], Diana A. Gelashvili [4], Yuri A. Shchepanovsky [4], Ivan A. Mizin [5], Danila V. Panchenko [6], Mikhail G. Bondar [7] and Taras P. Sipko [1]

1 Aleksey Nikolaevich Sevetrsov Institute of Ecology and Evolution of the Russian Academy of Sciences, 119071 Moscow, Russia; s_e_spiridonov@rambler.ru (S.E.S.); jolie15@yandex.ru (I.K.P.); sipkotp@mail.ru (T.P.S.)
2 Moscow Zoo, 123242 Moscow, Russia
3 Saint Petersburg State University of Veterinary Medicine, 196084 Saint Petersburg, Russia; larissabelova2010@yandex.ru (L.M.B.); fish2017@yandex.ru (Y.E.K.); darya.chuprak@mail.ru (D.I.C.)
4 Leningrad Zoo, 197198 Saint Petersburg, Russia
5 Nikolai Pavlovich Laverov Center for Integrated Arctic Research of the Ural Branch of the Russian Academy of Sciences, 163020 Arkhangelsk, Russia; ivan_mizin@mail.ru
6 Institute of Biology of Karelian Research Centre of the Russian Academy of Sciences, 185910 Petrozavodsk, Russia; danja@inbox.ru
7 Joint Directorate of Taimyr Nature Reserves, 663305 Norilsk, Russia; mikisayan@yandex.ru
* Correspondence: loginova_spb@bk.ru; Tel.: +7-(950)-029-5437

Abstract: Zoo conditions are unique for reindeer, since domestic reindeer are not kept in captive facilities like cattle. In the zoo, reindeer are usually surrounded by many different animals that they would never encounter naturally. Thus, they might be infected with new helminths. Numerous petting zoos raise concerns about the safety of tactile interactions for human visitors. Our study is the first large-scale one. Qualitative and quantitative fecal analyses were carried out for 233 reindeer distributed over 50 Russian zoos according to the National Standard of the Russian Federation (GOST R 54627-2011) *Ruminant animals—Methods of Laboratory Helminthological Diagnostics*. Where possible, DNA analyses of helminths were performed targeting internal transcribed spacer region. As a result, *F. hepatica*, *Paramphistomum* sp., *Moniezia* sp. (including *M. expansa*), gastrointestinal strongylids (including *Nematodirus* spp.), *Dictyocaulus* sp., *E. rangiferi*, *Trichuris* sp., and *Capillaria* sp. were found in 106 (45%) zoo reindeer. All these helminths were previously reported in reindeer and pose no direct danger for humans. The intensity of invasions was mostly low. Fecal examination might be considered as an indirect method for mange diagnostics, as *Chorioptes* and *Demodex* mites were found in reindeer fecal samples. The latter may represent a novel species of mite specific for reindeer.

Keywords: reindeer; *Rangifer tarandus*; helminth; trematode; cestode; nematode; mite; zoo

1. Introduction

Helminths in reindeer (*Rangifer tarandus*) have already been studied quite thoroughly, and the knowledge about them was summarized in thematic books [1–3] and chapters [4–8]. Ongoing studies expand our views on many aspects of reindeer helminths [9–13], and sometimes even add new species to the checklist of parasitic worms of *Rangifer* [14,15]. Helminths may cause nutritional deficiency in reindeer, decrease their physical condition, reduce their reproductive success, weaken immune system, cause anemia and blood loss, behavioral changes, damage to organs and tissues, and contribute to secondary infections [3–8,16].

Reindeer exist in domestic and wild forms, totaling more than 4 million animals worldwide [17]. Their popularity as zoo cervids in the 21st century is unprecedentedly growing. According to Species360 ZIMS database, there are more than 500 *Rangifer* spp. distributed in more than 150 zoos around the world [18]. Some research studies on reindeer helminths in captive facilities were published [3,16,19–21], but they are still too rare and fragmentary.

Meanwhile, zoo conditions are unique for reindeer because even domestic reindeer are not typically kept in farms or in captive facilities like cattle. In the zoo, reindeer are usually surrounded by many different animals that they would never encounter naturally, which might pose a risk of infection by new parasites [22,23]. Numerous petting zoos raise concerns about another issue: the safety of tactile interactions for human visitors.

Therefore, the aim of our study was to conduct a large-scale survey of reindeer helminths in Russian zoos. Fecal examination is a non-invasive method (an essential trait for zoo animals) that offers wide diagnostic capabilities, scoping gastrointestinal, lung, brain, and muscle parasitic worms. This study was carried out to find out what kinds of helminths inhabit zoo reindeer across Russia, how widespread they are, how typical they are, and how dangerous they are for both reindeer and humans.

2. Materials and Methods

2.1. Study Area

Our study collected fecal samples from 233 reindeer kept in 50 selected zoos and menageries located across Russia (Figure A1 and Table A1). Those include: all the state zoos in which collections reindeer are present (17); other state organizations that have live reindeer for exhibition (Children's ecological station, Natural-ethnographic complex, and Arctic Tourism Center); private organizations embers of the *Union of Zoos and Aquariums of Russia* (5 private zoos and an urban farm); and 25 private menageries. The latter constitute "clubs", "parks" ("recreation", "ethnic", "fauna", and "safari park"), "reserve" (self-proclaimed), "mini zoo", "ranch", "farm", "farmstead", "pets' corner" and unnamed menageries. Many of these organizations participated in our research anonymously; therefore, for uniformity, we do not provide the actual names of all the 50 of them. However, their location combined with the type of ownership makes the data quite transparent. For brevity, all of these organizations will be referred to as zoos.

2.2. Fecal Analysis

All fecal samples were processed according to the methods described in the National Standard of the Russian Federation (GOST R 54627-2011) *Ruminant animals—Methods of Laboratory Helminthological Diagnostics* [24]. Namely, they were the larvoscopic Vajda's method, combined ovoscopic Darling's method, and sedimentation ovoscopic method. The Vajda's method involves placing a few fecal pellets on a slide; to pour approximately 1 mL of 40 °C tap water on the pellets; to expose it for 30 min; to put away the pellets; and to examine the liquid microscopically. Flotation with Darling's solution involves mixing 3 g of feces with 60 mL of tap water; to filter and centrifuge the mixture; to pour off supernatant; to add the 1:1 mixture of glycerin and saturated sodium chloride solution (Darling's solution); to centrifuge the mixture; and to examine the supernatant microscopically. Sedimentation involves mixing 3 g of feces with 60 mL of tap water; to filter the mixture; to let it settle for 5 min; to pour off sediment; to resuspend it with water; to repeat until transparent supernatant; and to examine the sediment microscopically.

2.3. Helminths Identification

Primary identification of helminths at the stage of egg or larva was based on their morphology and morphometric data. The morphology of eggs and larvae derived from zoo reindeer feces was examined via light microscopy (LM) using the optical microscope Micmed-6 (LOMO-MA, Saint Petersburg, Russia) using the lenses of 4× (to navigate the slide), 10×, 20×, 40× magnification. The parasites were mounted between the slide and coverslip (24 × 24 mm). The larvae were in tap water. Eggs were either in tap water (once obtained via sedimentation) or in Darling's solution (derived via Darling's method). No specific dye to visualize the internal structures of parasites was used. Micrographs were taken using the digital photo camera 5D Mark II (Canon, Tokyo, Japan) connected to the microscope with the C-mount adapter (LOMO-MA, Russia). Morphometry was based on the obtained micrographs using Fiji/ImageJ version 1.2.4 RRID:SCR_003070 software (National Institutes of Health, Bethesda, MD, USA) set using the microscope calibration slide (transmitted light object micrometer) OMP (LOMO-MA, Russia). The Straight Line mode was used to measure eggs, and the Segmented Line mode was used to measure larvae. Reference literature was used to identify the obtained helminths [1–6,25]. Where possible, morphological identification was supported by DNA analysis.

2.4. DNA Analysis

For trematodes, the QIAamp DNA Accessory Set, Micro kit (Quiagen, Venlo, the Netherlands) was used to extract DNA from embryonated eggs as described by Loginova et al. [26]. For nematodes, DNA was extracted either via digestion of first larval stages (L1) in presence of Proteinase K [27] as described by Loginova et al. [28], or by using the QIAamp DNA Accessory Set, Micro kit (Quiagen, Venlo, Netherlands). We targeted the region of the internal transcribed spacer (ITS rDNA). For trematodes, we used BD1, BD2 primers [29] and corresponding protocols [30]. For nematodes we used either NC1, NC2 primers and protocols as described by Gasser et al. [31], or 18S, 26S primers [32] and protocols as described by Loginova et al. [28]. Visual check of PCR, cleaning of the PCR products, sequencing, and analysis of obtained chromatograms were performed as described by Loginova et al. [33].

2.5. Statistical Analysis

In this study, the statistical sample equals population, because all possible zoo reindeer were examined. A total of 233 zoo reindeer may not be representative of a larger group of reindeer. There are four breeds of domestic reindeer and few ecotypes or subspecies of wild reindeer in Russia. All these kinds of reindeer differ between each other, in particular in terms of helminth resistance. Zoo reindeer are heterogeneous. Some of them are known to originate from wild forest ecotype reindeer, others originated from domestic animals (but it is hard to identify their breed), and some have no remaining origin information. Regular reindeer exchanges between zoos make it even more complicated. Moreover, the above mentioned GOST R 54627-2011 prescribes to examine no less than 10% of animals. For approximately 2 million reindeer in Russia, the statistical sample must be around 200 thousand reindeer, which is almost 900 times larger than the sample size in our study. Thus, a 95% confidence interval (CI) to express the uncertainty of sample prevalence as an estimate of population prevalence is not applicable here. Therefore, to analyze the obtained data, the index method was used. Namely, it was prevalence and intensity. Sample prevalence was calculated as the proportion of positive samples (i.e., infected individuals) within the sample set (i.e., host sample) and expressed as a percentage (from 0–100%). The intensity for every reindeer was calculated as the number of parasites (eggs or larvae) per 1 g of their feces. The VIGIS chamber (analogue of the McMaster device) of the Diapar kit (VIGIS, Moscow, Russia) was used to calculate the intensity.

3. Results

Flukes, tapeworms and round worms were found at the stage of egg or larva in the feces of 106 reindeer (45%) kept in 31 zoos (62%). Voucher and GenBank accession numbers of helminths recovered from zoo reindeer feces are presented in Table 1.

The appearance of eggs and larvae of helminths obtained from zoo reindeer feces is shown in Figure 1.

Prevalence rates and distribution of helminths obtained from zoo reindeer feces are summarized in Table 2. For brevity, only helminth-positive zoos are included.

Prevalence rates for all 233 reindeer examined are shown in Figure 2.

Table 1. Species of helminths recovered from the feces of zoo reindeer and identified genetically.

Zoo ID	Species	GenBank [1]	Vouchers [2]
13	*Fasciola hepatica* [3]	PP328913	IPEE_Parasites 14320
4	*Elaphostrongylus rangiferi* [4]	PP843608	IPEE_Parasites 14319
5	*Elaphostrongylus rangiferi* [4]	PP843600	IPEE_Parasites 14339
6	*Elaphostrongylus rangiferi* [4]	PP843598	IPEE_Parasites 14343
7	*Elaphostrongylus rangiferi* [4]	PP843592	IPEE_Parasites 14344
8	*Elaphostrongylus rangiferi* [4]	PP845195	IPEE_Parasites 14345
9	*Elaphostrongylus rangiferi* [4]	PP845196	IPEE_Parasites 14346
12	*Elaphostrongylus rangiferi* [5]	MW848820	IPEE_Parasites 14282
13	*Elaphostrongylus rangiferi* [4]	PP843584	IPEE_Parasites 14347
25	*Elaphostrongylus rangiferi* [4]	PP845193	IPEE_Parasites 14348
29	*Elaphostrongylus rangiferi* [4]	PP845194	IPEE_Parasites 14349
30	*Elaphostrongylus rangiferi* [4]	PP845192	IPEE_Parasites 14350

[1] GenBank accession numbers for sequences from 65 embryonated eggs of *F. hepatica* and sets of 30 first larval stages (L1) of *E. rangiferi*. [2] Voucher specimens with definitive identifications and accession numbers archived in the Museum of Helminthological Collections of the Parasitology Center at the A. N. Severtsov Institute of Ecology and Evolution of the Russian Academy of Sciences (Moscow, Russia). [3] Sequence previously in GenBank as reported by Loginova et al. [26]. [4] Sequence information reported here for the first time. [5] Sequence previously in GenBank as reported by Loginova et al. [28].

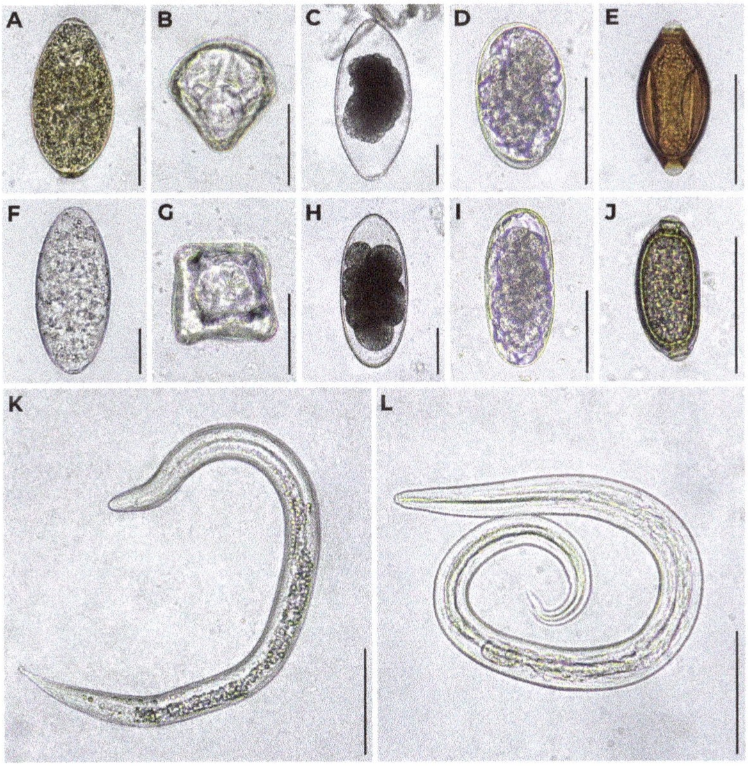

Figure 1. Diagnostic stages of helminths obtained from feces of zoo reindeer. (**A**) *Fasciola hepatica* egg; (**B**) *Moniezia expansa* egg; (**C**) *Nematodirus* sp. egg at late stage of embryonic development; (**D**) rounded strongyle-type egg; (**E**) *Trichuris* sp. egg; (**F**) *Paramphistomum* sp. egg; (**G**) *Moniezia* sp. egg; (**H**) *Nematodirus* sp. egg at early stage of embryonic development; (**I**) elongated strongyle-type egg; (**J**) *Capillaria* sp. egg; (**K**) *Dictyocaulus* sp. L1; (**L**) *Elaphostrongylus rangiferi* L1. Bright field microscopy, 400× magnification. Scale bar equals 50 μm.

Table 2. Helminths found in the feces of zoo reindeer. Prevalence was calculated for each zoo.

Zoo ID	Trematodes		Cestodes	Nematodes					
	Fasciola hepatica	*Paramphistomum* sp.	*Moniezia* spp.	Strongyle-Type	*Nematodirus* spp.	*Dictyocaulus* sp.	*Elaphostrongylus rangiferi*	*Trichuris* sp.	*Capillaria* sp.
2	–	–	–	–	–	–	–	–	4 (33%)
4	–	2 (18%)	–	–	–	–	4 (36%)	–	2 (18%)
5	–	2 (67%)	–	–	–	–	3 (100%)	–	–
6	–	–	–	–	–	–	1 (100%)	–	–
7	–	–	–	3 (43%)	–	–	4 (57%)	–	1 (14%)
8	–	–	–	–	–	–	1 (33%)	–	1 (33%)
9	–	3 (33%)	1 (11%)	–	–	–	6 (67%)	–	1 (11%)
10	–	–	–	–	–	–	–	–	1 (100%)
11	–	4 (50%)	–	–	–	–	–	–	–
12	–	–	–	4 (37%)	2 (18%)	–	7 (63%)	–	5 (45%)
13	1 (33%)	2 (67%)	–	–	–	–	1 (33%)	–	–
14	–	–	1 (50%)	2 (100%)	–	–	–	1 (50%)	–
16	–	–	–	–	–	–	–	–	2 (50%)
18	–	–	–	–	–	–	–	–	1 (50%)
21	2 (100%)	–	–	–	–	–	–	–	–
22	–	–	–	3 (100%)	–	–	–	1 (33%)	–
24	–	–	1 (17%)	1 (17%)	–	–	–	–	–
25	–	–	–	–	–	–	1 (17%)	–	1 (17%)
26	–	–	1 (25%)	–	–	–	–	–	2 (50%)
28	–	–	–	–	–	–	–	–	1 (20%)
29	–	–	–	–	–	–	2 (67%)	–	–
30	–	8 (38%)	–	1 (5%)	–	1 (5%)	6 (29%)	–	3 (14%)
34	–	–	1 (9%)	2 (18%)	–	–	–	–	2 (18%)
36	–	–	–	–	2 (29%)	–	4 (57%)	–	–
37	–	1 (33%)	–	–	–	–	3 (100%)	–	–
38	–	2 (100%)	–	2 (100%)	–	–	–	–	–
39	–	4 (80%)	–	–	–	–	–	–	–
44	–	–	–	–	1 (20%)	–	–	–	–
46	–	–	–	–	–	–	5 (100%)	–	–
47	–	–	–	–	–	–	–	–	1 (17%)
48	–	–	–	–	–	–	–	–	2 (100%)

Comparative intensity of infestation in the different zoos is presented in Table 3.

Table 3. Intensity of infestation with helminths in zoo reindeer according to GOST R 54627-2011.

Helminths	Intensity of Infestation Depending on the Number of Detected Helminth Eggs and Larvae, Specimens per 1 g of Feces			
	Low	Medium	High	Very High
Nematodes, Cestodes	**1–100**	**101–500** [1]	**501–1000**	**>1000**
GINs (eggs)	1–56 (zoos #7, 12, 14, 22, 24, 30, 34, 39)	–	–	–
Nematodirus spp. (eggs)	1–4 (zoos #12, 36, 44)	–	–	–
Dictyocaulus sp. (L1)	1 (zoo #30)	–	–	–
E. rangiferi (L1)	1–98 (zoos #4, 5, 6, 7, 8, 12, 13, 25, 29, 30, 36, 37, 46)	15–537 (zoo #9)	–	–
Trichuris sp. (eggs)	1 (zoos #14, 22)	–	–	–

Table 3. *Cont.*

Helminths	Intensity of Infestation Depending on the Number of Detected Helminth Eggs and Larvae, Specimens per 1 g of Feces			
	Low	**Medium**	**High**	**Very High**
Nematodes, Cestodes	**1–100**	**101–500** [1]	**501–1000**	**>1000**
Capillaria sp. (eggs)	1–8 (zoos #2, 4, 7, 8, 9, 10, 12, 16, 18, 25, 26, 28, 30, 34, 47, 48)	–	–	–
Moniezia spp. (eggs) [2]	1–17 (zoos # 9,14, 24, 26, 34)	–	–	–
Trematodes	**1–10**	**11–100**	**>100**	**-**
F. hepatica (eggs)	1–3 (zoos #13, 21)		–	–
Paramphistomum sp. (eggs)	1–8 (zoos #4, 5, 9, 11, 13, 37, 38, 38)	9–56 (zoo #30)	–	–

[1] Low intensity in the case of adult animals. [2] Including *M. expansa*.

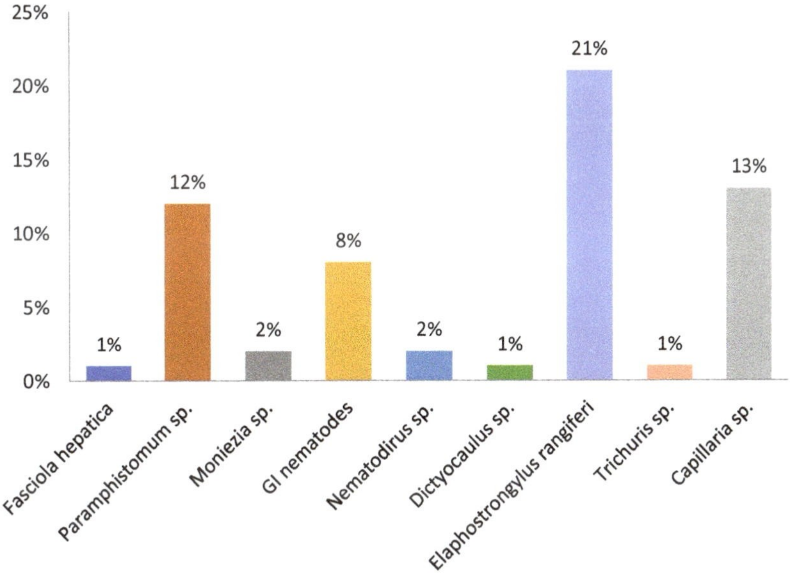

Figure 2. Prevalence rates (%) of helminths found in feces of zoo reindeer.

4. Discussion

Our study revealed the presence of trematodes (*Fasciola hepatica* and *Paramphistomum* sp.), cestodes (*Moniezia expansa* and *Moniezia* sp.), and nematodes (so-called small strongylids, *Nematodirus* spp., *Dictyocaulus* sp., *Elaphostrongylus rangiferi*, *Trichuris* sp., and *Capillaria* sp.) in zoo reindeer via fecal examination. All of these helminths were previously reported in reindeer [3,8].

In many samples, only one kind of helminth was found. However, we have also discovered singular combinations of two or three kinds of helminths (Table 4).

Common liver fluke, *F. hepatica*, inhabits the bile ducts of its host. Previously, it was reported in reindeer only in Eurasia, with one of the cases involving zoo reindeer [3]. This liver fluke seldom parasitizes reindeer, which is considered an accidental host. Infestation of reindeer with *F. hepatica* is usually associated with contact with other infected ruminants in captive facilities, where definitive hosts shed *F. hepatica* eggs that eventually develop into metacercaria [3]. In the two private zoos where *F. hepatica* was found in reindeer, sheep, goats and sika deer (*Cervus nippon*) were also infected.

Table 4. Combinations of helminths found in the feces of zoo reindeer.

Helminth 1	Helminth 2	Helminth 3
Fasciola hepatica	*Paramphistomum* sp.	–
Paramphistomum sp.	Small strongylids	–
Paramphistomum sp.	*Dictyocaulus* sp.	–
Paramphistomum sp.	*Elaphostrongylus rangiferi*	–
Paramphistomum sp.	*Elaphostrongylus rangiferi*	*Capillaria* sp.
Moniezia expansa	*Moniezia* sp.	–
Moniezia sp.	Small strongylids	–
Moniezia sp.	*Elaphostrongylus rangiferi*	–
Moniezia sp.	*Trichuris* sp.	–
Moniezia sp.	*Capillaria* sp.	–
Small strongylids	*Elaphostrongylus rangiferi*	–
Small strongylids	*Trichuris* sp.	–
Small strongylids	*Capillaria* sp.	–
Nematodirus sp. type 1	*Nematodirus* sp. type 2	–
Elaphostrongylus rangiferi	*Nematodirus* sp.	–
Elaphostrongylus rangiferi	*Capillaria* sp.	–

Rumen fluke, *Paramphistomum* sp., is widespread in Russian reindeer. When numerous, they cause atrophy of the rumen's papillae and may lead to death [3,8]. Reindeer from the private zoo #30, which had a medium level of invasion with rumen flukes, died within 6 months.

The tapeworms found were present with *M. expansa* (easily recognizable by its triangular eggs) [25] and *Moniezia* sp. These helminths inhabit the small intestine and may be quite pathogenic, especially for calves [3,8]. However, in our study, it was only found in adults. The reindeer is considered an accidental host for *M. expansa*, and its infestation is also associated with sympatric livestock and captive facilities [3].

Supposedly, the small strongylids found in zoo reindeer belong to two different genera (Figure 1D,I) which may belong to fam. Chabertiidae, Cooperiidae, Haemonchidae or Trichostrongylidae. Coproculture is needed to obtain larvae and clarify the diagnosis. In any case, small strongylids inhabit the gastrointestinal tract, and their impact depends on many circumstances [8,10].

The eggs of *Nematodirus* obtained from the reindeer fecal sample from zoo #40 certainly belonged to two different species. Apart from their differences in size and shape, their development rates were also different. Figure 1H shows *Nematodirus* sp. with eight blastomeres, whereas Figure 1C shows *Nematodirus* sp. with gastrulated embryo. These two types of eggs were found in the same fecal sample. In total, four eggs (two per type) were obtained from that sample. We attempted to obtain third larval stages (L3s) for DNA analysis but did not succeed. However, light micrographs (taken daily for three weeks after the finding of these eggs) showed their embryonic development and are available as Supplementary Materials. Helminths of *Nematodirus* sp. inhabit the small intestine. In reindeer, they are often accompanied by *Nematodirella longissimespiculata* [3], but in this study the latter was not found.

The lung worm *Dictyocaulus* sp. is the most concerning finding, because its presence may be life threatening. In the only zoo (#30) where it was found, there was also a sika deer infected with *Dictyocaulus* sp. It is hard to tell which of the animals was the source of invasion, or if it was it a coincidental infestation.

Brain worm, *E. rangiferi,* was the most widespread helminth in our research. Its pathogenicity is tricky because it may cause lethal epizootics or have no manifestation of its presence [3,8,34]. Due to the localization of adult worms in the nervous system or between the muscles, there is no effective treatment. On the other hand, authors observed a few reindeer that were severely infected in 2018 that eventually outlived their brain worms. At least no *E. rangiferi* larvae have been found in their fecal samples since 2023. Probably, it would be impossible in the wild, but in human care, reindeer can live much longer and, apparently, rid themselves of some parasites naturally.

Whipworms, *Trichuris* sp., inhabit the large intestine and can cause hemorrhagic diarrhea [8]. They are more frequent in captive animals. There are reindeer specific species and those common to ruminants that can also infect reindeer [3]. Egg morphology is not sufficient for the identification up to the species level.

Roundworm, *Capillaria* sp., inhabits the intestine and is often associated with young animals [3,8]. However, the impact of this parasite is not known. Infection with *Capillaria* sp. was previously reported in zoo reindeer as well [3].

Thus, via fecal examination, we managed to find liver and rumen flukes, GINs, lung worms, brain worms, whipworms, and capillarids. However, this method is not suitable for those helminths that inhabit ligaments (like *Onchocerca*), abdominal cavity (like *Setaria*), muscles, lungs, liver and brain (like larval stages of Taeniidae). Therefore, despite the large scale of our study, the helminth status of zoo reindeer is still incomplete.

Another bias of this work concerns different seasons of fecal sampling (which probably affected the reproductive activity of the helminths) and differences in age and sex of the studied animals [3,8]. These circumstances should be kept in mind when comparing the obtained results between zoos.

As we said earlier, the popularity of reindeer as zoo animals in the 21st century is outstanding. Only one zoo (#17) has a 100-year history of keeping reindeer in captivity. It is not surprising that fecal samples from their reindeer were helminth negative not only in this study, but years prior to it. Many other zoos obtained their first reindeer only in 2023. The majority of private zoos purchase domestic reindeer from agricultural husbandries. These husbandries are all private and run by families of indigenous people from the north. Zoos are able to buy only those reindeer that indigenous people are willing to sell. Reindeer herders want to keep the best individuals in the herd, and reindeer for sale may be less resistant to helminths.

The domestic reindeer, as opposed to wild reindeer, can be pure white (leucism, not albinism) throughout the year [35,36]. White reindeer attract most people [37], and numerous private zoos tend to have exactly the white reindeer. However, reindeer herders noticed that white reindeer are not good survivors [37,38]. It might add another disadvantage to such zoo reindeer in terms of helminth resistance and requires special research.

Apart from husbandry "heritage", zoo reindeer can become infected with helminths due to birds. Helminths eggs were reported to survive the sparrow digestive system and remain invasive [39]. Thus, birds can be vectors for helminths within the zoo and outside of its boundaries.

It is noteworthy that even in helminth-positive zoos, about 30% (from 0 to 83%) of reindeer fecal samples were helminth free, probably due to the different immune status of the reindeer.

Besides helminths, we found mites in reindeer fecal samples (Figure A2). They were either ingested during grooming and passed through the digestive tract or contaminated the fecal samples afterwards. Having found a mite egg in the sample from zoo #30, we contacted the zoo and shared our considerations. It turned out that the zoo indeed had problems with mange caused by mites (Figure A2E). One adult mite found in a sample from zoo #40 was identified as *Chorioptes* sp. (Figure A2D). *Chorioptes* mites were reported in reindeer in Canada, Norway, and Finland [8]. Another adult mite found in a reindeer fecal sample (zoo #12) met the morphological criteria for *Demodex* sp. To the authors' knowledge, *Demodex* has never been reported in reindeer [3,8,40]. This finding also requires additional research, which may result in the description of a new species.

Among the helminths found in the zoo reindeer in our research, there are a few genera reported in humans. Those are *Fasciola*, GINs (*Trichostrongylus*, *Oesophagostomum*), *Trichuris*, and *Capillaria* [41–44]. However, infected reindeer pose no direct threat for human visitors or zoo employees for the following reasons: the trematode *F. hepatica* is indeed a species that can infect both cervids and humans, but its life cycle requires an intermediate host (freshwater mollusk) [3]. The eggs shed by reindeer are not at infective stage for any vertebrate host. Moreover, eggs are excreted without miracidium, which takes time to develop (and is possible only in water). Species of *Trichostrongylus*, *Oesophagostomum*, *Trichuris*, and *Capillaria* that can infect humans are not the same species that infect reindeer [3,8,42,44]. Furthermore, all these nematodes shed eggs that are not embryonated by the time they are excreted in feces. Larvae development also requires a few days or weeks. Thus, if reindeer dung is regularly removed, even reindeer infected with helminths are safe for humans.

5. Conclusions

In our study, *F. hepatica*, *Paramphistomum* sp., *Moniezia* spp. (including *M. expansa*), gastrointestinal strongylids (including *Nematodirus* spp.), *Dictyocaulus* sp., *E. rangiferi*, *Trichuris* sp., and *Capillaria* sp. were found in 106 Russian zoo reindeer (out of 233) via fecal examination. All these helminths were previously reported in reindeer and pose no direct danger for humans. The intensity of invasions was mostly low. Fecal examination might be considered as an indirect method for mange diagnostics, as *Chorioptes* and *Demodex* mites were found in reindeer fecal samples. The latter may represent a novel species of mite specific for reindeer.

Supplementary Materials: The following supporting information can be downloaded from: https://www.mdpi.com/article/10.3390/jzbg5030033/s1, Figures S1–S25: Light microscopic images of the two types of *Nematodirus* eggs showing the development of embryos from 13 December 2023 to 4 January 2024, respectively. Eggs were obtained from the same fecal sample from a reindeer from zoo #40. The eggs were placed in a watch glass filled with tap water. Pictures were taken via photo camera using the smart phone Xperia2 DS Black (H4113) (Sony, China) and the optical microscope Micmed-6 (LOMO-MA, Russia).

Author Contributions: Conceptualization, O.A.L. and S.V.A.; methodology, O.A.L., D.N.E., S.E.S. and L.M.B.; formal analysis, O.A.L.; investigation, O.A.L. and S.E.S.; resources, O.A.L., S.V.A., D.N.E., N.S.E., S.E.S., I.K.P., L.M.B., Y.E.K., D.I.C., A.A.K., I.V.V., D.A.G., Y.A.S., I.A.M., D.V.P., M.G.B. and T.P.S.; writing—original draft preparation, O.A.L.; writing—review and editing, S.V.A., D.N.E., N.S.E., S.E.S., I.K.P., L.M.B., Y.E.K., D.I.C., A.A.K., I.V.V., D.A.G., Y.A.S., I.A.M., D.V.P., M.G.B. and T.P.S.; visualization, O.A.L.; supervision, L.M.B. and S.E.S.; project administration, S.V.A.; funding acquisition, I.A.M. All authors have read and agreed to the published version of the manuscript.

Funding: Part of this article was prepared within the framework of the State assignment for the N. Laverov Federal Center for Integrated Arctic Research of the Ural Branch of the Russian Academy of Sciences "Scientific foundations and social and cultural factors for the conservation and use of the potential of biological diversity in the European North and the Arctic" (registration number—122011400382-8)". Another part of this study was performed under the State order of IB KRC RAS (project FMEN-2022–0003).

Institutional Review Board Statement: Not applicable.

Data Availability Statement: Data are contained within the article.

Acknowledgments: Authors are deeply grateful to all the zoo administrators, veterinarians, keepers, and workers (as well as those not involved in zoo work directly) who assisted with fecal sampling and made this assistance possible. In particular, we would like to thank Alexander Makhov (Moscow), Kakhramon Khuzhanov (Moscow), Nelly Khafizova (Moscow), Andrey Soludkov (Moscow), Igor Tsaryov (Moscow), Andrey Kajdalov (Saint Petersburg), Lyudmila Zakharova (Saint Petersburg), Kristina Zabarina (Samara), Nataliya Chervyakova (Moscow), Vera Shiryaeva (Saint Petersburg), Irina Pavlova (Saint Petersburg), Daria Zimina (Izhevsk), Elena Kolesnikova (Khabarovsk), Vitaliy Soroka (Saint Petersburg), Irina Chuprak (Saint Petersburg), Irina Suvorova (Moscow), Evgeniy Popov and Alexander Sokolov from the Veterinary service of the Yamalo-Nenets Autonomous Okrug (Salekhard), and Vladimir Vasilev from the Northern Forum (Yakutsk). We also extend our gratitude to our colleagues who assisted in mite identification: Sergey Mironov from the Zoological Institute of the Russian Academy of Sciences (Saint Petersburg) and Nadezhda Gavrilova from the Saint

Petersburg State University of Veterinary Medicine (Saint Petersburg). Special thanks to Dmitry Kuznetsov for his suggestion to write this article.

Conflicts of Interest: The authors declare no conflicts of interest.

Appendix A

The study area is represented in Figure A1.

Figure A1. Map of Russia indicating the zoos. Insert on the right represents magnified area marked with a rectangle. Numbers correspond to the zoo ID numbers in Table A1.

Table A1. Collection data for fecal samples from reindeer (*Rangifer tarandus*) in the Russian zoos (federal subjects arranged from west to east).

Zoo ID	Ownership Type (S/P) [1]	Location (Federal Subject of Russia)	Coordinates (Decimal Degrees)	Number of Fecal Samples [3]	Date Collected
1	P	Murmansk Oblast	68.98716 33.07261	2	November 2023
2	P	Murmansk Oblast	68.85722 33.19556	12	November 2023
3	P	Murmansk Oblast	67.56317 33.36571	2	November 2023
4	P	Murmansk Oblast	67.65276 33.66051	11	November 2023
5	P	Murmansk Oblast	69.16835 35.13280	3	November 2023
6	P	Republic of Karelia	65.76340 31.07419	1	March 2024
7	P	Republic of Karelia	66.43669 32.85459	7	June 2022
8	P	Republic of Karelia	62.33297 34.00604	3	March 2024
9	P	Republic of Karelia	61.87838 34.07819	9	March 2024

Table A1. *Cont.*

Zoo ID	Ownership Type (S/P) [1]	Location (Federal Subject of Russia)	Coordinates (Decimal Degrees)	Number of Fecal Samples [3]	Date Collected
10	P	Leningrad Oblast	60.59067 30.00422	1	July 2020
11	P [2]	Leningrad Oblast	60.59161 30.11204	8	May 2019
12	P	Leningrad Oblast	60.14195 30.32823	11	August 2018
13	P	Leningrad Oblast	59.94762 30.68122	3	May 2023
14	P	Saint Petersburg	59.84108 30.06470	2	June 2019
15	P	Saint Petersburg	59.98017 30.24438	3	March 2024
16	P	Saint Petersburg	59.97054 30.25672	4	January 2020
17	S [2]	Saint Petersburg	59.95210 30.30891	9	March 2024
18	P	Saint Petersburg	59.67605 30.42401	2	February 2018
19	P	Tver Oblast	56.75051 36.36719	1	June 2020
20	S [2]	Moscow Oblast	55.94012 36.21055	7	January 2024
21	P	Moscow Oblast	56.13343 36.50051	2	March 2024
22	S [2]	Moscow	55.76347 37.57537	3	December 2023
23	P [2]	Moscow	55.83306 37.62197	1	October 2023
24	S [2]	Yaroslavl Oblast	57.67709 39.90005	6	November 2023
25	P [2]	Nizhny Novgorod Oblast	56.33468 43.85420	6	November 2023
26	P	Nizhny Novgorod Oblast	56.92384 45.40185	4	August 2022
27	S [2]	Republic of Mordovia	54.17500 45.18599	3	November 2023
28	S [2]	Vologda Oblast	60.74784 46.17739	5	November 2023
29	P [2]	Ulyanovsk Oblast	54.35485 48.52409	3	February 2024
30	P	Samara Oblast	53.34493 50.22240	21	August 2021
31	S [2]	Udmurt Republic	56.86555 53.17413	2	November 2023
32	S	Nenets Autonomous Okrug	67.63436 53.24135	1	February 2024

Table A1. *Cont.*

Zoo ID	Ownership Type (S/P) [1]	Location (Federal Subject of Russia)	Coordinates (Decimal Degrees)	Number of Fecal Samples [3]	Date Collected
33	S [2]	Perm Krai	58.01672 56.23728	4	November 2023
34	P	Sverdlovsk Oblast	57.17585 60.65764	11	February 2024
35	S [2]	Chelyabinsk Oblast	55.16894 61.36764	5	November 2023
36	P	Tyumen Oblast	56.99540 65.73485	7	February 2024
37	S	Yamalo-Nenets Autonomous Okrug	66.59257 66.85846	3	January 2024
38	S [2]	Yamalo-Nenets Autonomous Okrug	66.07485 76.65427	2	November 2023
39	S [2]	Omsk Oblast	56.08978 74.64219	5	November 2023
40	S [2]	Novosibirsk Oblast	55.05612 82.88010	7	December 2023
41	P [2]	Altai Krai	53.35593 83.68230	2	November 2023
42	S [2]	Tomsk Oblast	56.60427 84.86807	1	March 2024
43	P	Krasnoyarsk Krai	69.42091 88.26126	1	February 2024
44	S [2]	Krasnoyarsk Krai	55.96669 92.73100	5	November 2023
45	S [2]	Republic of Sakha (Yakutia)	61.67818 129.35184	5	April 2024
46	P	Republic of Sakha (Yakutia)	62.03243 129.72416	5	August 2020
47	S [2]	Khabarovsk Krai	48.62218 135.06819	6	December 2019
48	S [2]	Sakhalin Oblast	46.96788 142.75403	2	December 2019
49	S [2]	Kamchatka Krai	53.18850 158.38604	1	April 2019
50	P	Kamchatka Krai	55.92095 158.69500	3	March 2024

[1] S is for State ownership, P is for Private ownership. [2] Member of the *Union of Zoos and Aquariums of Russia*.
[3] Number of fecal samples equals the number of reindeer in the zoo.

Appendix B

Mites found in the feces of zoo reindeer and manifestation of mange caused by mites in reindeer is shown in Figure A2.

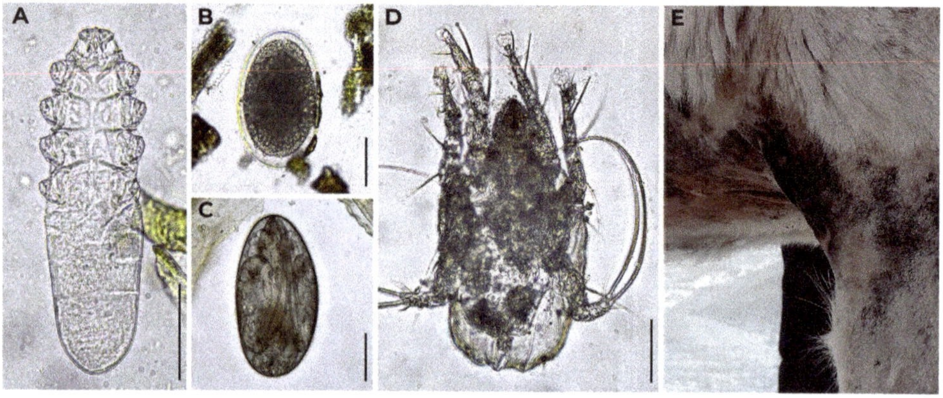

Figure A2. Mites of reindeer. (**A**) *Demodex* sp. obtained from the feces of a reindeer from zoo #12; (**B**) egg of mite at early developmental stage obtained from the feces of a reindeer from zoo #30; (**C**) egg of mite at late developmental stage obtained from the feces of a reindeer from zoo #48; (**D**) *Chorioptes* sp. obtained from the feces of a reindeer from zoo #40; (**E**) manifestation of mange caused by mites in a reindeer from zoo #30 (furless patches in the front leg and chest are shown); photo courtesy: Kristina Zabarina. Light micrographs were made via bright field microscopy, 400× magnification. Scale bar equals 50 μm.

References

1. Skrjabin, K.I. *Glistnye Invazii Severnogo Olenya (Helminth Diseases of Reindeer)*; Selkhozgiz: Moscow, Russia, 1931; 88p. (In Russian)
2. Polyanskaya, M.V. *Gel'mintosy Severnykh Oleney (Helminthiases of Reindeer)*; Knizhnoe Izdatelstvo: Murmansk, Russia, 1963; 47p. (In Russian)
3. Mizkewitsch, V.Y. *Gel'minty Severnogo Olenya i Vyzyvayemyye imi Zabolevaniya (Reindeer Helminths and the Diseases They Cause)*; Kolos: Saint Petersburg, Russia, 1967; 308p. (In Russian)
4. Golosov, I.M.; Mizkewitsch, V.Y. Helminthiases. In *Parazitarnye Bolezni Severnykh Oleney (Parasitic Diseases of Reindeer)*; Lyzhin, K., Ed.; Krasnoyarskoe Knizhnoe Izdatelstvo: Krasnoyarsk, Russia, 1964; pp. 57–141. (In Russian)
5. Laaksonen, S. Trematoda. Cestoda. Nematoda. In *TUNNE PORO. Poron Sairaudet ja Terveydenhoito (FEEL THE REINDEER. Reindeer Diseases and Health Care)*; Livonia Print: Riga, Latvia, 2016; pp. 220–251. (In Finnish)
6. Belova, L.M.; Loginova, O.A. Helminthiases. In *Bolezni Severnykh Oleney (Diseases of Reindeer)*; Zabrodin, V.A., Laishev, K.A., Eds.; Knizhnoe Izdatelstvo: Saint Petersburg, Russia, 2019; pp. 94–113. (In Russian)
7. Kutz, S.J.; Ducrocq, J.; Verocai, G.G.; Hoar, B.M.; Colwell, D.D.; Beckmen, K.B.; Polley, L.; Elkin, B.T.; Hoberg, E.P. Parasites in ungulates of Arctic North America and Greenland: A view of contemporary diversity, ecology, and impact in a world under change. *Adv. Parasitol.* **2012**, *79*, 99–252. [CrossRef] [PubMed]
8. Kutz, S.J.; Laaksonen, S.; Asbakk, K.; Nilssen, A.C. Helminths. In *Reindeer and Caribou. Health and Disease*; Tryland, M., Kutz, S.J., Eds.; CRC Press: Boca Raton, FL, USA, 2019; pp. 180–205.
9. Hrabok, J.T.; Oksanen, A.; Nieminen, M.; Waller, P.J. Population dynamics of nematode parasites of reindeer in the sub-arctic. *Vet. Parasitol.* **2006**, *142*, 301–311. [CrossRef] [PubMed]
10. Jokelainen, P.; Moroni, B.; Hoberg, E.; Oksanen, A.; Laaksonen, S. Gastrointestinal parasites in reindeer (*Rangifer tarandus tarandus*): A review focusing on Fennoscandia. *Vet. Parasitol. Reg. Stud. Rep.* **2019**, *17*, 100317. [CrossRef] [PubMed]
11. Emelyanova, A.; Savolainen, A.; Oksanen, A.; Nieminen, P.; Loginova, O.; Abass, K.; Rautio, A. Research on Selected Wildlife Infections in the Circumpolar Arctic—A Bibliometric Review. *Int. J. Environ. Res. Public Health* **2022**, *19*, 11260. [CrossRef] [PubMed]
12. Dickinson, E.R.; Orsel, K.; Cuyler, C.; Kutz, S.J. Life history matters: Differential effects of abomasal parasites on caribou fitness. *Int. J. Parasitol.* **2023**, *53*, 221–231. [CrossRef] [PubMed]
13. Andrew, C.L.; Wagner, B.; Harms, N.J.; Jenkins, E.J.; Jung, T.S. Comparative Prevalence and Intensity of Endoparasites in a Dynamic Boreal Ungulate Community. *Diversity* **2024**, *16*, 230. [CrossRef]
14. Verocai, G.G.; Kutz, S.J.; Simard, M.; Hoberg, E.P. *Varestrongylus eleguneniensis* sp. n. (Nematoda: Protostrongylidae): A widespread, multi-host lungworm of wild North American ungulates, with an emended diagnosis for the genus and explorations of biogeography. *Parasites Vectors* **2014**, *7*, 1–22. [CrossRef]
15. Loginova, O.A.; Kolpashchikov, L.A.; Spiridonov, S.E. First report of *Orthostrongylus* sp. (Nematoda: Protostrongylidae) in wild reindeer (*Rangifer tarandus*) from the Taimyr, Russia: Nearctic parasites in a Palearctic host. *Parasitol. Res.* **2023**, *122*, 685–689. [CrossRef] [PubMed]

16. Hoberg, E.P.; Kocan, A.A.; Rickard, L.G. Gastrointestinal strongyles in wild ruminants. In *Parasitic Diseases of Wild Mammals*; Samuel, W.M., Pybus, M.J., Kocan, A.A., Eds.; Wiley: New York, NY, USA, 2001; pp. 193–227.
17. Tryland, M.; Kutz, S.J. *Reindeer and Caribou. Health and Disease*; CRC Press: Boca Raton, FL, USA; London, UK; New York, NY, USA, 2019; p. 3, ISBN 978-1-4822-5068-8.
18. Species 360. Data Science for Zoos and Aquariums. Available online: https://species360.org/products-services/zoo-aquarium-animal-management-software-2/ (accessed on 31 May 2024).
19. Haigh, J.C.; Mackintosh, C.; Griffin, F. Viral, parasitic and prion diseases of farmed deer and bison. *Rev. Sci. Tech.* **2002**, *21*, 219–248. [CrossRef]
20. Flach, E. Cervidae and Tragulidae. In *Zoo and Wild Animal Medicine*, 5th ed.; Fowler, M.E., Miller, E.R., Eds.; Elsevier Science: Amsterdam, The Netherlands, 2003; pp. 634–649.
21. Goossens, E.; Vercruysse, J.; Boomker, J.; Vercammen, F.; Dorny, P. A 12-month survey of gastrointestinal helminth infections of cervids kept in two zoos in Belgium. *J. Zoo Wildl. Med.* **2005**, *36*, 470–478. [CrossRef]
22. Manninen, S.M.; Thamsborg, S.M.; Laaksonen, S.; Oksanen, A. The reindeer abomasal nematode (*Ostertagia gruehneri*) is naturally transmitted to sheep when sharing pastures. *Parasitol. Res.* **2014**, *113*, 4033–4038. [CrossRef] [PubMed]
23. Utaaker, K.S.; Ytrehus, B.; Davey, M.L.; Fossøy, F.; Davidson, R.K.; Miller, A.L.; Robertsen, P.-A.; Strand, O.; Rauset, G.R. Parasite spillover from domestic sheep to wild reindeer—The role of salt licks. *Pathogens* **2023**, *12*, 186. [CrossRef]
24. Latypov, D.G.; Timerbayeva, R.R.; Kirillov, E.G. *Parazitologiya i Invazionnyye Bolezni Zhvachnykh Zhivotnykh (Parasitology and Invasive Diseases of Ruminants)*; Lan: Saint Petersburg, Russia, 2019; pp. 59–67, ISBN 978-5-8114-3561-6. (In Russian)
25. Verocai, G.G.; Chaudhry, U.N.; Lejeune, M. Diagnostic Methods for Detecting Internal Parasites of Livestock. *Vet. Clin. Food. Anim.* **2020**, *36*, 125–143. [CrossRef]
26. Loginova, O.; Efeykin, B.; Krutikova, A.; Mizin, I.; Spiridonov, S. *Fasciola hepatica*: Updates on egg morphology, host range, and distribution. *Food Waterborne Parasitol.* **2024**, *36*, e00237. [CrossRef]
27. Holterman, M.; van der Wurff, A.; van den Elsen, S.; van Megen, H.; Bongers, T.; Holovachov, O.; Bakker, J.; Helder, J. Phylum-Wide Analysis of SSU rDNA Reveals Deep Phylogenetic Relationships among Nematodes and Accelerated Evolution toward Crown Clades. *Mol. Biol. Evol.* **2006**, *23*, 1792–1800. [CrossRef] [PubMed]
28. Loginova, O.A.; Belova, L.M.; Spiridonov, S.E. The First Report on *Elaphostrongylus rangiferi* (Reindeer Invasive Parasite) in Leningrad Oblast. *Russ. J. Biol. Invasions* **2022**, *13*, 232–244. [CrossRef]
29. Luton, K.; Walker, D.; Blair, D. Comparisons of ribosomal internal transcribed spacers from two congeneric species of flukes (Platyhelminthes: Trematoda: Digenea). *Mol. Biochem. Parasitol.* **1992**, *56*, 323–327. [CrossRef] [PubMed]
30. Mahami-Oskouei, M.; Dalimi, A.; Forouzandeh-Moghadam, M.; Rokni, M.B. Molecular identification and differentiation of *Fasciola* isolates using PCR-RFLP method based on internal transcribed spacer (ITS1, 5.8 S rDNA, ITS2). *Iranian J. Parasitol.* **2011**, *6*, 35–42.
31. Gasser, R.B.; Chilton, N.B.; Hoste, H.; Beveridge, I. Rapid sequencing of rDNA from single worms and eggs of parasitic helminths. *Nucleic Acids Res.* **1993**, *21*, 2525–2526. [CrossRef]
32. Vrain, T.C.; Wakarchuk, D.A.; Levesque, A.C.; Hamilton, R.I. Intraspecific rDNA restriction fragment length polymorphism in the *Xiphinema americanum* group. *Fundam. Appl. Nematol.* **1992**, *15*, 563–573.
33. Loginova, O.A.; Rozenfeld, S.B.; Sipko, T.P.; Mizin, I.A.; Panchenko, D.V.; Laishev, K.A.; Bondar, M.G.; Kolpashchikov, L.A.; Gruzdev, A.R.; Kulemeev, P.S.; et al. Diversity and Distribution of Helminths in Wild Ruminants of the Russian Arctic: Reindeer (*Rangifer tarandus*), Muskoxen (*Ovibos moschatus*), and Snow Sheep (*Ovis nivicola*). *Diversity* **2023**, *15*, 672. [CrossRef]
34. Handeland, K.; Slettbakk, T. Outbreaks of Clinical Cerebrospinal Elaphostrongylosis in Reindeer (*Rangifer tarandus tarandus*) in Finnmark, Norway, and their Relation to Climatic Conditions. *J. Vet. Med. Ser. B* **1994**, *41*, 407–410. [CrossRef]
35. Baskin, L.M. *Odomashnivaniye Severnogo Olenya. Ot Okhotnika do Pastukha i Ranchevoda (Reindeer Domestication. From Hunter to Herdsman and Rancher)*; Tovarishchestvo Nauchnykh Izdaniy KMK: Moscow, Russia, 2021; p. 280, ISBN 978-5-907372-89-4. (In Russian)
36. Holand, Ø.; Mizin, I.; Weladji, R.B. Reindeer *Rangifer tarandus* (Linnaeus, 1758). In *Handbook of the Mammals of Europe*; Hackländer, K., Zachos, F.E., Eds.; Springer: Cham, Switzerland, 2022; pp. 247–277, ISBN 978-3-030-24474-3. [CrossRef]
37. Smith, T. *Velvet Antlers, Velvet Nose. The Story of a Reindeer Family*; Coronet Books, Hodder and Stoughton: London, UK, 1995; p. 90, ISBN 0-340-66003-1.
38. Wella, Y. *Azbuka Olenevoda (Reindeer Herder's ABC)*; Studiya O.K.: Surgut, Russia, 2011; 43p. (In Russian)
39. Mosgovoy, A.A. Sparrows as agent distributing helminthic infections among domestic animals. In *Papers on Helminthology*; Schultz, R.-E.S., Gnedina, M.P., Eds.; All-Union Lenin Academy of Agricultural Sciences: Moscow, Russia, 1937; pp. 398–402. (In Russian)
40. Izdebska, J.N.; Fryderyk, S. *Demodex acutipes* Bukva et Preisler, 1988 (Acari, Demodecidae)—A rare parasite of red deer (*Cervus elaphus* L.). *Ann. Parasitol.* **2012**, *58*, 161–166. [PubMed]
41. Vasilevich, F.I.; Belova, L.M.; Burmistrova, M.I. *Parazitarnye Zoonozy (Parasitic Zoonoses)*; ZooVetKniga: Moscow, Russia, 2020; pp. 21–28, ISBN 978-5-6045650-5-6. (In Russian)
42. Mas-Coma, S.; Bargues, M.D.; Valero, M.A. Human fascioliasis infection sources, their diversity, incidence factors, analytical methods and prevention measures. *Parasitology* **2018**, *145*, 1665–1699. [CrossRef]

43. Lattes, S.; Ferte, H.; Delaunay, P.; Depaquit, J.; Vassallo, M.; Vittier, M.; Kokcha, S.; Coulibaly, E.; Marty, P. *Trichostrongylus colubriformis* Nematode Infections in Humans, France. *Emerg. Inf. Dis.* **2011**, *17*, 1301–1302. [CrossRef]
44. Dhaliwal, B.B.S.; Juyal, P.D. *Textbook of Parasitic Zoonoses*; Springer Nature: Singapore, 2022; pp. 46–110, ISBN 978-81-322-1550-9.

Journal of
Zoological and Botanical Gardens

MDPI

Article

Gastrointestinal Parasitic Infections in Non-Human Primates at Gabon's Primatology Center: Implications for Zoonotic Diseases

Krista Mapagha-Boundoukou [1,*], Mohamed Hassani Mohamed-Djawad [1,2], Neil Michel Longo-Pendy [1], Patrice Makouloutou-Nzassi [1,3], Félicien Banguebouss a [1], Mourad Ben Said [4,5], Barthélémy Ngoubangoye [3,6] and Larson Boundenga [1,7,*]

1 Health Ecology Research Unit (URES), Interdisciplinary Centre for Medical Research of Franceville (CIRMF), Franceville BP 769, Gabon; mhdjawad2020@gmail.com (M.H.M.-D.); longo2michel@gmail.com (N.M.L.-P.); patmak741@gmail.com (P.M.-N.); bangueboussafelicien040@gmail.com (F.B.)
2 Laboratory of Evolutionary Biology, Ecology and Ecosystem Management, Department of Animal Biology, Faculty of Science and Technology, Cheikh Anta Diop University, Dakar BP 5005, Senegal
3 Department of Animal Biology and Ecology, Tropical Ecology Research Institute (IRET/CENAREST), Libreville BP 13354, Gabon; genistha@hotmail.com
4 Department of Basic Sciences, Higher Institute of Biotechnology of Sidi Thabet, University of Manouba, Manouba 2010, Tunisia; bensaidmourad83@yahoo.fr
5 Laboratory of Microbiology, National School of Veterinary Medicine of Sidi Thabet, University of Manouba, Manouba 2010, Tunisia
6 Primatology Center, Interdisciplinary Centre for Medical Research of Franceville (CIRMF), Franceville BP 769, Gabon
7 Anthropology Department, University of Durham, South Road, Durham DH1 3LE, UK
* Correspondence: mapaghakrista@gmail.com (K.M.-B.); boundenga@gmail.com (L.B.)

Abstract: Parasites and infectious diseases pose significant threats to primate populations, especially in captive non-human primates (NHPs). This study aimed to assess the diversity and prevalence of intestinal parasites in NHPs at the CIRMF Primatology Center. A total of 97 fecal samples were analyzed using parasitological techniques, including sodium chloride flotation and modified Baermann sedimentation methods. An overall parasite prevalence of 93.81% (91/97) was observed. Sixteen groups of parasites with zoonotic potential were identified, comprising ten genera of nematodes (*Trichuris*, *Enterobius*, Hookworm, *Trichostrongylus*, *Mammomonogamus*, Spirure, *Oesophagostomum*, *Schistosoma*, *Ascaris*, and *Strongyloides*), three genera of protists (*Eimeria*, *Balantioides coli/Buxtonella*, and *Entamoeba*), one genus of cestodes (*Hymenolepis*), and two genera of trematodes (*Dicrocoelium* and *Paramphistomum*). High prevalences were noted for *Oesophagostomum* spp. (83.5%), *Strongyloides* spp. (52.58%), and *Trichostrongylus* spp. (50.52%). These findings underscore the potential role of the CIRMF Primatology Center in maintaining and facilitating the transmission of intestinal parasites with high zoonotic potential. The co-existence of human and NHP parasites in shared environments, such as zoos and research facilities, emphasizes the need for a holistic, One Health approach that addresses the interconnected health of humans, animals, and the environment. This study highlights the urgent need for collaborative strategies to mitigate the risks of zoonotic parasite transmission between NHPs and humans in captive settings.

Keywords: non-human primates; gastrointestinal parasites; prevalence; zoonotic diseases; CIRMF primatology center; one health approach

1. Introduction

Parasitic diseases pose a significant global health challenge [1–4]. Approximately 1.5 billion people are infected with intestinal parasites annually, leading to approximately 135,000 deaths [5]. This statistic highlights the critical need to address parasitic infections from both human and zoonotic perspectives. The close genetic similarity between humans and great apes, coupled with increased interactions through activities such as hunting, agriculture, logging, ecotourism, urbanization, and the domestication of non-human primates (NHPs) significantly elevates the risk of cross-species transmission of parasites [6–8].

Great apes infected with intestinal parasites not only face serious health issues like malnutrition, anemia, and secondary infections but also represent a threat to human populations [1,9]. These health impairments directly affect their survival and reproductive success, exacerbating the conservation challenges faced by already endangered great ape populations [10,11]. In settings like the Primatology Center, where humans and various NHPs species interact closely, the risk of pathogen exchange is significantly heightened [12], unlike in the wild where natural barriers reduce such interactions [13].

The Primatology Center of the Interdisciplinary Center for Medical Research in Franceville (CIRMF) and Gabonese sanctuaries are dedicated to the care and rehabilitation of orphaned NHPs and those rescued from illegal private ownership [14–16]. A dedicated team of caretakers, veterinarians, and nurses work tirelessly to ensure the proper nutrition and overall welfare of these animals. Despite these efforts, NHPs at the center remain susceptible to a range of infections, including viral, fungal, bacterial, and parasitic diseases [17–21]. Among these, gastrointestinal parasites (GIPs) are particularly prevalent posing significant health risks such as growth retardation, gastrointestinal disorders, abortions, and neurological problems [22,23].

While the establishment of protected areas, sanctuaries, and primatology centers are essentials for NHP conservation [24,25], these confined environments also facilitate the zoonotic and anthropozoonotic transmission of pathogens [1,26–28]. The SARS-CoV-2 pandemic serves as a stark reminder of the risks posed by close human–animal interactions. Given the ongoing decline in great ape population [29], the CIRMF Primatology Center plays a critical role in understanding the impact of gastrointestinal parasites on NHP health and their conservation. This study aims to assess the diversity of intestinal parasites at the center, highlighting their potential effects on both primate and human health. By doing so, it seeks to provide valuable insights that can inform and improve conservation efforts, ultimately contributing to the long-term survival of these endangered species.

2. Materials and Methods

2.1. Study Site and Sample Collection

This study was conducted at the Primatology Center of CIRMF, located in southeastern Gabon (Figure 1). The center features enclosures with natural ground for free-ranging animals as well as aviaries with cemented floors. Sampling collection took place between February to April 2023, from 9 a.m to 1 p.m. A total of approximately 97 fecal samples were collected from five (5) primate species. Table 1 shows the distribution of the NHP species sampled, their characteristics, and status under the International Union for Conservation of Nature (UICN).

Figure 1. Location of the primatology center.

Table 1. Distribution of the population according to each sampled monkey species.

Common Name	Species	Habitat	Male	Female	UICN Status	Protection Status in Gabon	Total
Macaque	*Macaca rhesus*	aviary	7	3	last concern 2015	-	10
Nictitans	*Cercopithecus nictitans*	aviary	1	3	endangered 2020	-	4
Chimpanzee	*Pan t. troglodytes*	aviary	14	11	endangered 2016	fully protected	25
Mandrill	*Mandrillus sphinx*	enclosure	20	20	vulnerable 2016	fully protected	40
Solatus	*Allochrocebus solatus*	enclosure	2	16	near threatened 2019	fully protected	18
Total			44	53			97

To ensure the precise and uncontaminated fecal collection from NHPs, animals were sequentially captured in the feeding zone. Each individual was identified by their tattoo or ear tag. Fecal samples were collected immediately after defecation, avoiding ground contact, and placed in labeled coprology containers indicating species, sex, and collection date. Samples were either analyzed on the same day or stored at room temperature for a later analysis within 48 h. A microscopic examination of eggs and cysts was conducted using a Leica DM2000 LED microscope equipped with a Leica DFC450 digital camera for image capture.

2.2. Microscopic Analysis

Fecal samples were processed immediately using flotation and sedimentation methods as previously described [30,31]. An average of 2 g of fecal matter was used. This amount was added to a saline solution, the concentration of which varied according to flotation or sedimentation (40% and 9%, respectively). The staining step with bromothymol was omitted to enhance the visualization of parasite eggs and oocysts. Larval forms were extracted using the Baermann method. Parasite identification was based on morphological characteristics, color, and content, following the guidelines of [32,33]. In this study, to differentiate between *Necator* and *Ancylostoma* eggs, we focused on the biological behavior of the eggs. Notably, *Ancylostoma* eggs typically hatch within 24 h post-emission [34,35]. Thus, any eggs identified 48 h after collection were classified as *Necator* spp. However, as this distinction is still uncertain, we decided to refer to the worms *Ancylostoma* and *Necator* as 'hookworms'. Although larval forms belonging to the *Strongyloides* and Enterobius genera have been observed, it was not possible to identify the corresponding species precisely. To assess the parasitic load, the protocol by [36] was followed, using 2 g of fecal matter. The calculation of the parasitic load was calculated using the following formula:

$$\text{EPG} = (\text{Total number of eggs counted}/\text{Number of grids counted}) \times (\text{Total volume (mL)}/\text{Examined volume (mL)} \times 50\ (\text{Dilution factor}).$$

2.3. Statistical Analysis

The data for this study were analyzed using R software (version 4.3.0). The prevalence of a given parasite was calculated as the ratio of the number of individuals positive for that parasite to the total number of individuals examined. To compare the prevalence of parasitic infections among different primate taxa, we employed the Kruskal–Wallis test, which is suitable for non-parametric continuous data. This test allowed us to determine if there were significant differences in infection rates between groups. Additionally, Fisher's exact test was applied to compare the number of infected and uninfected animals, providing a robust assessment of the impact of infections within various populations. The Shannon diversity index (H) and equitability (E) were also calculated to evaluate the parasitic diversity within the taxa, offering insights into the richness and evenness of the present parasitic species. Finally, a linear regression model was utilized to examine the influences of taxon, habitat, and sex on parasitic load, thereby identifying significant factors contributing to the variations in parasitic infections.

3. Results

3.1. Diversity and Distribution of Parasite Genera in Non-Human Primates at the CIRMF Primatology Center

This study identified 3 protists and 13 helminths in 5 examined NHP species (Figure 2 and Figure S1). All NHPs were infected with four common helminths: *Trichuris*, *Oesophagostomum*, *Trichostrongylus*, and *Strongyloides*. The distribution of the remaining nine helminths and three protists was as follows: Hookworm found in chimpanzee, macaque, mandrill, and solatus; *Schistosoma* in mandrill; *Mammomonogamus* in chimpanzee and mandrill; *Spirura* in macaque, mandrill, and solatus; *Enterobius* in chimpanzee, mandrill, and nictitans; *Hymenolepis* in solatus; *Dicrocoelium* in sacaque and solatus; *Paramphistomum* in solatus; *Balantioides coli*/*Buxtonella* in chimpanzee, macaque, mandrill, and solatus; *Eimeria* in mandrill; and *Entamoeba* in chimpanzee.

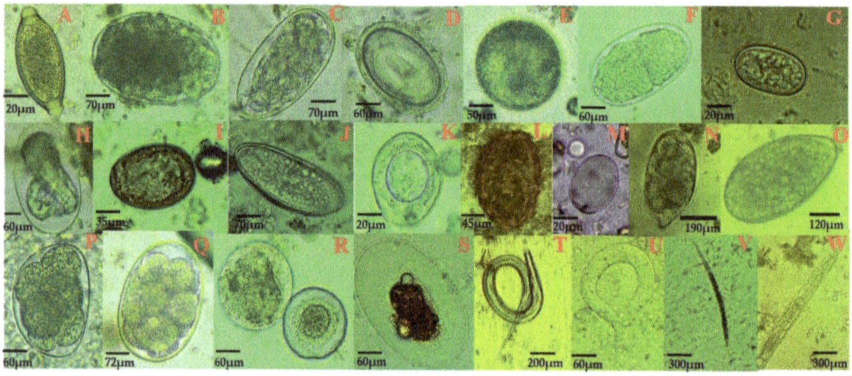

Figure 2. Parasitic structures identified in the feces of NHPs at the CIRMF Primatology Center (**A**), *Trichuris* sp. (**B**,**T**), *Oesophagostomum* sp. (**C**), *Trichostrongylus* sp. (**D**), Spirure (**E**), *Balantioides coli/Buxtonella* sp. (**F**), *Mammomonogamus* sp. (**G**), *Eimeria* sp. (**H**,**W**), *Strongyloides* sp. (**I**), *Dicrocelium* sp. (**J**,**U**), *Enterobius* sp. (**K**), *Hymenolepis* sp. (**L**), *Ascaris* sp. (**M**), *Entamoeba* sp. (**N**), *Schistosoma* sp. (**O**), *Paramphistomum* sp. (**P**,**V**,**Q**), Hookworm (**R**,**S**), unidentified eggs.

For all the parasites, species diversity, as reflected by Shannon's diversity index and equitability, indicated high parasite species diversity across all NHP species. However, the highest diversity (Shannon Index) and equitability were recorded in mandrills (H:2.92; E:1.81), followed by solatus (H:2.89; E:1.80), chimpanzee (H:2.47; E:1.53), and macaques (H:2.46; E:1.53) (Figure 3).

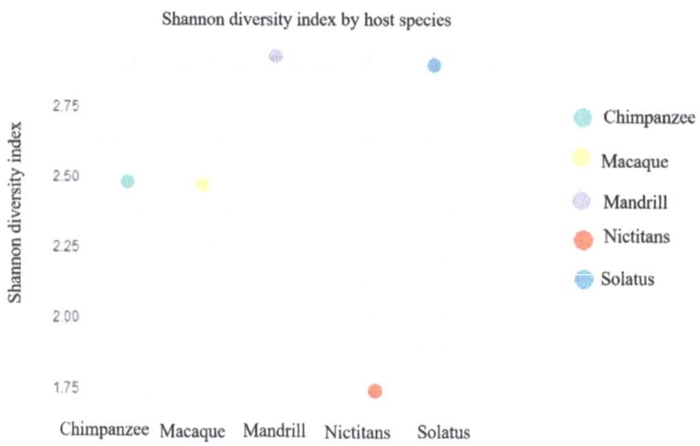

Figure 3. Shannon diversity index by host species.

3.2. Prevalence of Parasites in NHPs at CIRMF Primatology Center

The analysis of samples collected from non-human primates at the Primatology Center of CIRMF revealed an overall parasitic infection prevalence of 93.81%, providing an overview of the parasitic status within the studied population. The analysis of parasite prevalence revealed significant variations among different primate species, including chimpanzees, macaques, mandrills, nictitans, and solatus. In terms of overall prevalence, mandrills exhibited the highest rate (97.5%), followed by chimpanzees (96%), solatus (100%), nictitans (75%), and macaques (70%) (Table 2).

Table 2. Prevalence of parasite by host species.

Groups	Taxa	Chimpanzee	Macaque	Mandrill	Nictitans	Solatus	Means
Nematoda	*Trichuris* sp.	16	20	7.50	25	33.33	16.49
	Oesophagostomum sp.	88	50	90	50	88.89	83.50
	Trichostrongylus sp.	60	20	60	25	38.89	50.52
	Hookworm	33.33	20	32.5	0	24	29.03
	Schistosoma spp.	0	0	5	0	0	2.06
	Mammomonogamus sp.	4	0	2.5	0	0	2.06
	Strongyloides sp.	56	30	65	25	38.89	52.58
	Ascaris sp.	0	0	2.5	0	0	1.03
	Spirure	0	20	22.5	0	5.56	12.37
	Enterobius sp.	16	0	10.26	25	0	9.37
Cestoda	*Hymenolepis* sp.	0	0	0	0	11.11	2.06
Trematoda	*Dicrocelium* sp.	0	10	0	0	11.11	3.09
	Paramphistomum sp.	0	0	0	0	5.56	1.03
Protist	*Balantioides coli/Buxtonella* sp.	68	10	42.5	0	11.11	38.14
	Eimeria sp.	0	0	7.5	0	0	3.09
	Entamoeba sp.	8	0	0	0	0	2.06
Percentage (%) of animals with at least one parasite		96	70	97.5	75	100	-
p-value (Kruskal–Wallis)				0.406			

Upon closer examination, we also assessed the prevalence of each parasite species within each taxon (Table 2). Chimpanzees are distinguished by a particularly high prevalence of *Oesophagostomum* (88%) and *Balantioides coli/Buxtonella* (68%), indicating significant exposure to these parasites. In contrast, macaques show moderate prevalence levels for several parasites, including *Strongyloides* (30%) and Hookworm (20%). Mandrills also display high prevalence rates, particularly for *Oesophagostomum* (90%) and *Trichostrongylus* (60%). Nictitans exhibit varied prevalence, peaking at 25% for *Trichuris*. Finally, solatus are characterized by a notable prevalence of *Oesophagostomum* (88.89%) and *Strongyloides* (38.89%) (Table 2).

To identify the most prevalent parasite at the primatology center, we calculated the prevalence of each parasitic species in our study population sample without the distinction of taxon. *Oesophagostomum* spp. emerged as the most widespread parasite, infecting 83.51% of individuals. Other parasites, such as *Strongyloides* spp. (52.58%), *Trichostrongylus* spp. (50.52%), and *Balantioides coli/Buxtonella* (38.14%), were also detected at significant levels. Conversely, certain parasites, such as *Ascaris* spp. (1.03%) and *Mammomonogamus* spp. (2.06%), exhibited relatively low infection rates (Figure 4).

The following table illustrates the impact of gender, habitat, and host species on the degree of parasitism. The statistical analysis revealed no statistically significant differences ($p > 0.05$) in parasite infestation levels in relation to the factors under investigation. The infestation rate of females was similar to that of males (0.409), irrespective of habitat ($p = 0.082$) or host species ($p = 0.406$). The proportion of infected animals ranged from 70% to 100%, with the macaque group exhibiting the lowest level of infestation (50%), whereas all solatus were infected with at least one parasite (Table 3).

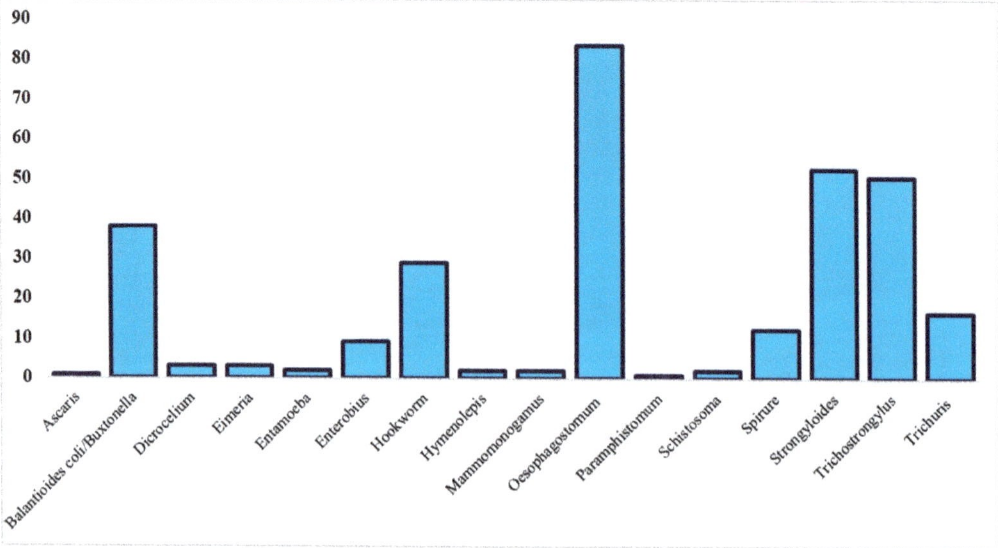

Figure 4. Prevalence of each parasite genera.

Table 3. Impact of gender and habitat factors on parasitism levels.

Factors	Classes	Sample (N)	Infected	Prevalence Rate (% ± Standard Deviation)	Df	*p*-Value
Living farm	Enclosure Aviary	57 40	56 35	98.25 ± 0.70 87.5 ± 3.54	1	0.082
Sex	Male Female	44 53	41 50	93.18 ± 2.12 94.33 ± 2.12	1	0.4096
Hosts	Chimpanzee Macaque Mandrill Nictitans Solatus	25 10 40 4 18	24 7 39 3 18	96 ± 0.70 70 ± 2.12 97.5 75 ± 0.70 100	4	0.406

3.3. *Influence of Analyzed Variables on Parasite Load*

The analysis of the linear regression model revealed significant influences of taxon and habitat on parasitic load.

The results indicate that mandrills and solatus exhibit significantly higher parasitic loads compared to chimpanzees, which serve as the reference group (Figure 5). Specifically, mandrills show a coefficient of 2510.2 ($p = 0.00698$), while solatus have a coefficient of 2139.9 ($p = 0.02517$). These findings suggest that these two taxa are more vulnerable to parasitic infections. Conversely, macaques display a significantly lower parasitic load, with a coefficient of −741.7 ($p = 0.02851$).

Regarding habitat, primates living in aviaries also demonstrate a high parasitic load, with a coefficient of 2882.6 ($p = 0.00198$). In contrast, the analysis did not reveal a significant effect of sex on parasitic load, as indicated by the coefficient of −322.4 ($p = 0.10323$) for males compared to females.

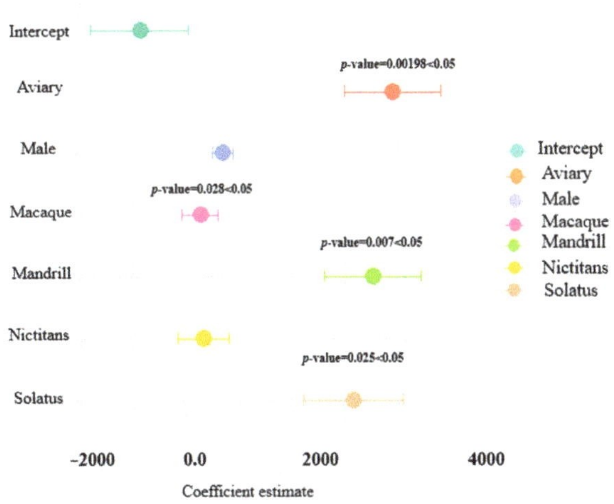

Figure 5. Influence of variables (hosts, sex, and species) on parasite load.

4. Discussion

Primatology centers are crucial for the understanding of parasite exchange dynamics, especially the transmission of parasites between primates and humans due to frequent interactions [27,37]. Our study aimed to elucidate the diversity of intestinal parasites in captive NHPs at CIRMF's Primatology Center. In the present study, we found a global parasite infestation prevalence of 93,81% (91/97) among the studied primates, which is notably higher than previous captive NHPs studied in Africa (67% to 76.2%) [18,38,39] and Asia (89.6%) [40]. This discrepancy may be attributed to a nearly two-year lapse in deworming at the CDP due to limited resources. This high prevalence highlights the increased susceptibility of NHPs to gastrointestinal parasites and remains consistent with the report, which reports a prevalence ranging from 22 to 100% in NHPs [15].

Our coprological analyses identified a total of 16 parasitic taxa infecting NHPs. The diversity of parasites observed varied across studies, largely due to the significant influence of environmental conditions on the dynamics of gastrointestinal parasites, as previously documented in the literature [15,41–43]. Since environmental factors differ from one location to another, they can substantially affect parasite diversity. Among the taxa identified in this study, some species belonging to *Oesophagostomum*, *Enterobius*, *Ascaris*, *Trichostrongylus*, *Strongyloides*, and hookworms are particularly known for their high zoonotic potential. Species of these genera have been implicated in bidirectional transmission in environments shared by NHPs and humans [44–46]. However, a reliance on microscopic analysis alone limits the ability to ensure accurate identification down to the species level. A notable instance is *Enterobius anthropopitheci*, the primary parasite of the *Enterobius* genus found in the chimpanzees [47]. This parasite is closely related to *Enterobius vermicularis* with which it shares similar morphological characteristics [48]. Documented instances of co-infection between these two species have been observed in environments where humans and primates coexist [49]. In light of this complexity, the utilization of molecular tools is imperative. These advanced techniques are of great importance for resolving the identification ambiguities associated with cryptic species and for gaining deeper insights into the dynamics of infection in areas of human–primate interaction. This finding underscores the need for rigorous feeding protocols to minimize pathogen exchange risks. Additionally, NHPs may act as reservoirs for human-infecting parasites, indicating that primatology centers could be sources of parasitic infections that pose significant health risks, particularly to young children [50].

The most prevalent parasitic genera identified were *Oesophagostomum* (83.5%), *Strongyloides* (52.58%), *Trichostrongylus* (50.52%), *Balantioides coli/Buxtonella* (38.14%), Hookworm (29. 03%), and *Trichuris* (16.49%). These findings are consistent with studies in sub-Saharan Africa, which often reported helminth infections as the most common among the NHP population [33,51,52]. The high prevalence of *Oesophagostomum* may be due to its low immunogenicity, allowing the effective colonization of host environments. Additionally, the promiscuity and reuse of the same soil in captive and semi-captive settings can facilitate the spread of intestinal parasites [12,18]. Nematodes from the *Oesophagostomum* genus are known to cause severe diseases in primates, including granulomas, caseous lesions, and abscesses in the intestinal wall, with some NHPs potentially acting as reservoirs for human oesophagostomosis [53].

The highest diversity and load of intestinal parasites were observed in mandrills, chimpanzees, and solatus, likely due to their direct contact with natural soil, which harbors a wider variety of parasites. Natural soil serves as a reservoir for infectious forms of intestinal parasites, promoting multi-parasitism in these animals [22,53,54]. Our study found similar infection rates in males and females, regardless of habitat or species, consistent with the findings of Eke et al. [55]. The absence of significant differences between sexes may be due to the communal living conditions of NHPs, where both males and females experience the same level of parasite exposure, as shown in other studies [55,56]. Differences in parasite infestation rates between sexes in NHPs are often associated with differences in home range and foraging behaviors. At our center, both sexes are housed together and receive equal food from caretakers [57,58]. Our results also show that individuals living in aviaries (cemented floor) had a much higher parasite burden than those living in pens (natural floor). Our observations are similar to those of Opeyemi et al. [59]. on helminth infections in captive birds, confirming the risks associated with aviary habitats. This finding may be explained by the fact that natural floors, in contrast to cemented ones, possess a capacity for self-regulation of parasite populations [59].

The potential for cross-species transmission, especially in captive settings with close contact between humans and NHPs, is a serious concern [18,60–63]. This issue would not only jeopardize the health of both primates and humans but also impacts the conservation of NHP populations. It is imperative to consider animal, human, and environmental factors to improve illness prevention at this interface. Effective measures should include improved hygiene, sanitation, and veterinary care.

The One Health approach is crucial to effectively addressing concerns related to intestinal parasites. This framework integrates efforts across human, animal, and environmental sectors, enhancing surveillance, improving environmental management, and fostering collaboration among stakeholders [6]. By recognizing the interconnectedness of these health domains, the One Health approach is essential for improving health outcomes and promoting sustainable coexistence between humans and NHPs. This holistic perspective is essential for combating intestinal parasites and advancing public health and conservation efforts.

5. Conclusions

The findings of this study highlight a significant risk of zoonotic disease transmission associated with the presence of gastrointestinal parasites in primatology centers in Gabon. The high prevalence and diversity of these parasites among non-human primates (NHPs) highlight the urgent need for enhanced surveillance and robust biosecurity measures. Our findings identified several genera of parasites, including *Strongyloides*, *Oesophagostomum*, Hookworm, and *Enterobius*, which harbor species with zoonotic potential that could present a risk to public health in cohabitation scenarios. Furthermore, these parasites may have a significant impact on the health of NHPs, potentially leading to increased morbidity and susceptibility to other infections. To gain a full understanding of the implications of these findings, further molecular analysis is essential to trace the origins and transmission pathways of these parasites. The close contact between humans and NHPs presents significant health risks for both groups, reinforcing the necessity of a One Health approach to ensure the safety of all parties. This integrative framework acknowledges the interconnectivity between human, animal, and environmental health, which is essential for the effective management of zoonotic risks. To mitigate these threats, it is essential to implement a systematic monitoring program for potential parasitic infections and to restrict unauthorized contact between visitors and staff with NHP food and water sources. By implementing these preventive measures and adopting a One Health perspective, we can protect the health of both primates and humans, fostering a safer coexistence in shared environments while enhancing our collective capacity to address zoonotic threats comprehensively.

Supplementary Materials: The following supporting information can be downloaded at: https://www.mdpi.com/article/10.3390/jzbg5040048/s1, Figure S1: Various forms of larvae of observed gastrointestinal parasites.

Author Contributions: Conceptualization, L.B. and M.B.S.; methodology, K.M.-B.; software, K.M.-B., M.H.M.-D. and N.M.L.-P.; validation, L.B., N.M.L.-P. and P.M.-N.; formal analysis, K.M.-B., F.B. and M.H.M.-D.; investigation, K.M.-B., F.B. and M.H.M.-D.; resources, L.B., K.M.-B. and B.N.; data curation, N.M.L.-P., F.B. and P.M.-N.; writing—original draft preparation, K.M.-B.; writing—review and editing, L.B., P.M.-N., N.M.L.-P., M.H.M.-D. and M.B.S.; visualization, N.M.L.-P., P.M.-N. and MBS; supervision, L.B.; project administration, L.B.; funding acquisition, L.B. All authors have read and agreed to the published version of the manuscript.

Funding: This research received no external funding.

Institutional Review Board Statement: This study was approved by the scientific committee of our institute, the Centre Interdisciplinaire de Recherches Medicales de Franceville (CIRMF), in accordance with the ethical principles of animal research. All samples were collected with due regard to animal welfare, and, in this study, all samples were collected with the consent of the animal owners. In addition, animal sampling procedures were evaluated by the Institutional Committee for Animal Use and Care of the National CIRMF.

Data Availability Statement: All data generated or analyzed during this study are included in this published article.

Acknowledgments: We would like to express our gratitude to all members of Primatology center, the interdisciplinary medical research center in Franceville, for their assistance in learning coprological techniques. We would also like to acknowledge the various small ruminant breeders with whom we collaborated, as well as the interdisciplinary medical research center in Franceville for their technical support.

Conflicts of Interest: The authors declare no conflicts of interest.

References

1. Medkour, H.; Amona, I.; Laidoudi, Y.; Davoust, B.; Bitam, I.; Levasseur, A.; Akiana, J.; Diatta, G.; Pacheco, L.; Gorsane, S.; et al. Parasitic Infections in African Humans and Non-Human Primates. *Pathogens* **2020**, *9*, 561. [CrossRef] [PubMed]
2. Altizer, S.; Nunn, C.L.; Lindenfors, P. Do threatened hosts have fewer parasites? A comparative study in primates. *J. Anim. Ecol.* **2007**, *76*, 304–314. [CrossRef] [PubMed]
3. El Kouni, M.; Chu, C. Purine metabolism in parasites: Potential targets for chemotherapy. In *Recent Advances in Nucleosides: Chemistry and Chemotherapy*; Chu, C.K., Ed.; Elsevier Science: Amsterdam, The Netherlands, 2002; pp. 377–416. [CrossRef]
4. Rondón, S.; Cavallero, S.; Di Filippo, M.M.; De Liberato, C.; Berrilli, F.; Capitani, N.; D'amelio, S. Intestinal parasites infecting captive non-human primates in Italy. *Front. Vet. Sci.* **2024**, *10*, 1270202. [CrossRef]
5. Organisation Mondiale de la Santé. Soil-Transmitted Helminth Infections. 2023. Available online: https://www.who.int/news-room/fact-sheets/detail/soil-transmitted-helminth-infections (accessed on 28 October 2024).
6. Boundenga, L.; Makouloutou-Nzassi, P.; Ngoubangoye, B. A review of Gabonese gorillas and their pathogens: Diversity, transfer and One Health approach to avoid future outbreaks? *Front. Parasitol.* **2023**, *2*, 1115316. [CrossRef]
7. Prugnolle, F.; Rougeron, V.; Becquart, P.; Berry, A.; Makanga, B.; Rahola, N.; Arnathau, C.; Ngoubangoye, B.; Menard, S.; Willaume, E.; et al. Diversity, host switching and evolution of *Plasmodium vivax* infecting African great apes. *Proc. Natl. Acad. Sci. USA* **2013**, *110*, 8123–8128. [CrossRef] [PubMed]
8. Jiang, X.; Fan, Z.; Li, S.; Yin, H. A Review on Zoonotic Pathogens Associated with Non-Human Primates: Understanding the Potential Threats to Humans. *Microorganisms* **2023**, *11*, 246. [CrossRef]

9. Obanda, V.; Maingi, N.; Muchemi, G.; Ng'ang'a, C.J.; Angelone, S.; Archie, E.A. Infection dynamics of gastrointestinal helminths in sympatric non-human primates, livestock and wild ruminants in Kenya. *PLoS ONE* **2019**, *14*, e0217929. [CrossRef]
10. Gillespie, T.R.; Lonsdorf, E.V.; Canfield, E.P.; Meyer, D.J.; Nadler, Y.; Raphael, J.; Pusey, A.E.; Pond, J.; Pauley, J.; Mlengeya, T.; et al. Demographic and ecological effects on patterns of parasitism in eastern chimpanzees (*Pan troglodytes schweinfurthii*) in Gombe National Park, Tanzania. *Am. J. Phys. Anthropol.* **2010**, *143*, 534–544. [CrossRef]
11. Hockings, K.J.; McLennan, M.R.; Carvalho, S.; Ancrenaz, M.; Bobe, R.; Byrne, R.W.; Dunbar, R.I.; Matsuzawa, T.; McGrew, W.C.; Williamson, E.A.; et al. Apes in the Anthropocene: Flexibility and survival. *Trends Ecol. Evol.* **2015**, *30*, 215–222. [CrossRef]
12. Boundenga, L.; Moussadji, C.; Mombo, I.M.; Ngoubangoye, B.; Lekana-Douki, J.B.; Hugot, J.-P. Diversity and prevalence of gastrointestinal parasites in two wild Galago species in Gabon. *Infect. Genet. Evol.* **2018**, *63*, 249–256. [CrossRef]
13. Daszak, P.; Cunningham, A.A.; Hyatt, A.D. Emerging infectious diseases of wildlife—Threats to biodiversity and human health. *Science* **2000**, *287*, 443–449. [CrossRef] [PubMed]
14. Sutherland, C.A.; Dicks, L.V.; Petrovan, S.O.; Smith, R.K. *What Works in Conservation. 2021*; Open Book Publishers: Cambridge, UK, 2020; p. 794.
15. Vonfeld, I.; Prenant, T.; Polack, B.; Guillot, J.; Quintard, B. Gastrointestinal parasites in non-human primates in zoological institutions in France. *Parasite* **2022**, *29*, 43. [CrossRef] [PubMed]
16. Calle, P.; Joslin, J.O. Singes du Nouveau Monde et singes de l'Ancien Monde. In *Fowler's Zoo and Wild Animal Medicine*; Elsevier: Amsterdam, The Netherlands, 2015; Volume 8, pp. 301–335.
17. Andrade, M.C.; Lemos, B.R.; Silva, L.M.; Pecotte, J.K. Eliminating potential effects of other infections during selection of nonhuman primates for COVID-19 research. *Comp. Med.* **2023**, *73*, 45–57. [CrossRef] [PubMed]
18. Boundenga, L.; Ngoubangoye, B.; Moukodoum, N.; Dibakou, S.-E.; Moussadji, C.; Hugot, J.P. Diversity of parasites in two captive chimpanzee populations in southern Gabon. *Infect. Genet. Evol.* **2021**, *91*, 104807. [CrossRef]
19. Ngoubangoye, B.; Boundenga, L.; Arnathau, C.; Mombo, I.M.; Durand, P.; Tsoumbou, T.-A.; Otoro, B.V.; Sana, R.; Okouga, A.-P.; Moukodoum, N.; et al. The host specificity of ape malaria parasites can be broken in confined environments. *Int. J. Parasitol.* **2016**, *46*, 737–744. [CrossRef] [PubMed]
20. Ngoubangoye, B.; Boundenga, L.; Dibakou, S.-E.; Tsoumbou, T.-A.; Kinga, C.M.; Prugnolle, F.; Fouchet, D.; Pontier, D. Surgical treatment of *Oesophagostomum* spp. nodular infection in a chimpanzee at the CIRMF primatology Center, Gabon. *Case Rep. Vet. Med.* **2021**, *2021*, 6617416. [CrossRef]
21. Ngoubangoye, B.; Fouchet, D.; Boundenga, L.A.; Cassan, C.; Arnathau, C.; Meugnier, H.; Tsoumbou, T.-A.; Dibakou, S.E.; Ekore, D.O.; Nguema, Y.O.; et al. *Staphylococcus aureus* host spectrum correlates with methicillin resistance in a multi-species ecosystem. *Microorganisms* **2023**, *11*, 393. [CrossRef]
22. Sirima, C.; Bizet, C.; Hamou, H.; Červená, B.; Lemarcis, T.; Esteban, A.; Peeters, M.; Ngole, E.M.; Mombo, I.M.; Liégeois, F.; et al. Soil-transmitted helminth infections in free-ranging non-human primates from Cameroon and Gabon. *Parasites Vectors* **2021**, *14*, 354. [CrossRef]
23. Nicolas, X.; Chevalier, B.; Simon, F.; Klotz, F. Traitement des parasitoses intestinales (amibiase et mycoses exclues). *Encycl. Méd. Chir.* **2002**, *1*, 1–14.
24. Brent, L. Life-long well being: Applying animal welfare science to nonhuman primates in sanctuaries. *J. Appl. Anim. Welf. Sci.* **2007**, *10*, 55–61. [CrossRef]
25. Feliu, O.; González-Zamora, A.; Riba, D.; Sauquet, T.; Sánchez-López, S.; Maté, C. The impact of sanctuary visits on children's knowledge and attitudes toward primate welfare and conservation. *PeerJ* **2023**, *11*, e15074. [CrossRef] [PubMed]
26. Senghore, M.; Bayliss, S.C.; Kwambana-Adams, B.A.; Foster-Nyarko, E.; Manneh, J.; Dione, M.; Badji, H.; Ebruke, C.; Doughty, E.L.; Thorpe, H.A.; et al. Transmission of *Staphylococcus aureus* from humans to green monkeys in the Gambia as revealed by whole-genome sequencing. *Appl. Environ. Microbiol.* **2016**, *82*, 5910–5917. [CrossRef] [PubMed]
27. Dawet, A.; Yakubu, D.; Butu, H. *Survey of Gastrointestnal Parasite of Non-Human Primates in Jos Zoological Garden*; University of Jos: Jos, Nigeria, 2013. Available online: https://dspace.unijos.edu.ng/jspui/bitstream/123456789/276/1/2167-6801-2-108(1).pdf (accessed on 28 October 2024).
28. Weingartl, H.M.; Embury-Hyatt, C.; Nfon, C.; Leung, A.; Smith, G.; Kobinger, G. Transmission of Ebola virus from pigs to non-human primates. *Sci. Rep.* **2012**, *2*, 811. [CrossRef] [PubMed]
29. Fritz, J. Using analytics to nudge student responsibility for learning. *New Dir. High. Educ.* **2017**, *2017*, 65–75. [CrossRef]
30. Lester, H.; Matthews, J. Faecal worm egg count analysis for targeting anthelmintic treatment in horses: Points to consider. *Equine Vet. J.* **2014**, *46*, 139–145. [CrossRef]
31. Dryden, M.W.; Payne, P.A.; Ridley, R.; Smith, V. Comparison of Common Fecal Flotation Techniques for the Recovery of Parasite Eggs and Oocysts. *Vet. Ther.* **2005**, *6*, 15–28.
32. Huffman, M.A.; Chapman, C.A. (Eds.) *Primate Parasite Ecology: The Dynamics and Study of Host-Parasite Relationships*; Cambridge University Press: Cambridge, UK; New York, NY, USA, 2009.
33. Roland, K.W.Y.; Sylvie, P.; Fidèle, B.K.; Hilaire, B.K. Parasites Gastro-intestinaux des populations humaines du parc national de Taï, Côte d'Ivoire. *Eur. Sci. J.* **2019**, *15*, 27–44. [CrossRef]
34. Soulsby, E.J.L. Helminths, Arthropods and Protozoa of Domesticated Animals. 1968. Available online: https://www.cabidigitallibrary.org/doi/full/10.5555/19682902735 (accessed on 27 October 2024).

35. Hotez, P.J.; Brooker, S.; Bethony, J.M.; Bottazzi, M.E.; Loukas, A.; Xiao, S. Hookworm infection. *N. Engl. J. Med.* **2004**, *351*, 799–807. [CrossRef]
36. Vasconcelos-Nóbrega, C.; Santos, C.; Mega, C.; Coelho, C.; Cruz, R.; Vala, H.; Esteves, F.; Mesquita, J.R. *ABC Series on Diagnostic Parasitology Part 2: The McMaster Method*; MA Healthcare Limited: London, UK, 2017.
37. Adetunji, V. Prevalence of gastro-intestinal parasites in primates and their keepers from two zoological gardens in Ibadan, Nigeria. *Sokoto J. Vet. Sci.* **2014**, *12*, 25–30. [CrossRef]
38. Mbaya, A.; Udendeye, U. Gastrointestinal parasites of captive and free-roaming primates at the Afi Mountain Primate Conservation Area in Calabar, Nigeria and their zoonotic implications. *Pak. J. Biol. Sci.* **2011**, *14*, 709–714. [CrossRef]
39. N'da, K.M.; Dahourou, L.D.; Gbati, O.B.; Alambedji, R.B. Diversity and prevalence of gastrointestinal parasites with zoonotic potential of Green Monkeys in Bandia Reserve in Senegal. *Int. J. One Health* **2021**, *7*, 65–69. [CrossRef]
40. Adrus, M.; Zainuddin, R.; Ahmad Khairi, N.H.; Ahamad, M.; Abdullah, M.T. Helminth parasites occurrence in wild proboscis monkeys (*Nasalis larvatus*), endemic primates to Borneo Island. *J. Med. Primatol.* **2019**, *48*, 357–363. [CrossRef] [PubMed]
41. Mbora, D.N.; McPeek, M.A. Host density and human activities mediate increased parasite prevalence and richness in primates threatened by habitat loss and fragmentation. *J. Anim. Ecol.* **2009**, *78*, 210–218. [CrossRef] [PubMed]
42. Stoner, K.E.; González Di Pierro, A.M. Intestinal parasitic infections in Alouatta pigra in tropical rainforest in Lacandona, Chiapas, Mexico: Implications for behavioral ecology and conservation. In *New Perspectives in the Study of Mesoamerican Primates: Distribution, Ecology, Behavior, and Conservation*; Springer: Berlin/Heidelberg, Germany, 2006; pp. 215–240.
43. Gillespie, T.R.; Nunn, C.L.; Leendertz, F.H. Integrative approaches to the study of primate infectious disease: Implications for biodiversity conservation and global health. *Am. J. Phys. Anthropol.* **2008**, *137*, 53–69. [CrossRef] [PubMed]
44. Ota, N.; Hasegawa, H.; McLennan, M.R.; Kooriyama, T.; Sato, H.; Pebsworth, P.A.; Huffman, M.A. Molecular identification of *Oesophagostomum* spp. from 'village' chimpanzees in Uganda and their phylogenetic relationship with those of other primates. *R. Soc. Open Sci.* **2015**, *2*, 150471. [CrossRef]
45. Laidoudi, Y.; Medkour, H.; Latrofa, M.S.; Davoust, B.; Diatta, G.; Sokhna, C.; Barciela, A.; Hernandez-Aguilar, R.A.; Raoult, D.; Otranto, D.; et al. Zoonotic *Abbreviata caucasica* in Wild Chimpanzees *(Pan troglodytes verus*) from Senegal. *Pathogens* **2020**, *9*, 517. [CrossRef]
46. Nosková, E.; Sambucci, K.M.; Petrželková, K.J.; Červená, B.; Modrý, D.; Pafčo, B. Strongyloides in non-human primates: Significance for public health control. *Philos. Trans. R. Soc. B* **2024**, *379*, 20230006.
47. Hasegawa hideo Modry, D.; Tenora, F.; Vallo, P. Two nez subspecies of *Enterobius* (Nematoda: Oxyuridae) fromthe chimpanzee (*Pan troglodyes*) in equatorial Africa. *J. Parasitol.* **1992**, *78*, 676–684.
48. Hugot, J.-P.; Reinhard, K.J.; Gardner, S.L.; Morand, S. Human enterobiasis in evolution: Origin, specificity and transmission. *Parasite* **1999**, *6*, 201–208. [CrossRef]
49. Ghandour, A.; Zahid, N.; Banaja, A.; Kamal, K.; Bouq, A. Zoonotic intestinal parasites of hamadryas baboons Papio hamadryas in the western and northern regions of Saudi Arabia. *J. Trop. Med. Hyg.* **1995**, *98*, 431–439.
50. Ghislaine, O.D. Prévalence des Parasitoses Intestinales chez Lesenfants de 0 à 5, Dans la Communauté D'anonkoi 3. Ph.D. Thesis, Felix Houphouët Boigny University, Abidjan, Ivory Coast, 2015.
51. Setchell, J.M.; Bedjabaga, I.-B.; Goossens, B.; Reed, P.; Wickings, E.J.; Knapp, L.A. Parasite Prevalence, Abundance, and Diversity in a Semi-free-ranging Colony of Mandrillus sphinx. *Int. J. Primatol.* **2007**, *28*, 1345–1362. [CrossRef]
52. Kouassi, A.F.; Aké-Assi, E.; N'goran, K.S.B.; Ouattara, D.; Tiebré, M.S. Contribution de l'élevage urbain à la sécurité alimentaire: Stratégies d'adaptation des éleveurs de bovins dans le District d'Abidjan, Côte d'Ivoire. *Afr. Sci.* **2019**, *15*, 218–228.
53. Guillot, J.; Vermeulen, B.; Lafosse, S.; Chauffour, S.; Cibot, M.; Narat, V.; Masi, S.; Nieguitsila, A.; Snounou, G.; Bain, O.; et al. Les nématodes du genre Oesophagostomum. Un risque émergent pour l'homme et les grands singes en Afrique? *Bull. L'académie Natl. Méd.* **2011**, *195*, 1955–1963. [CrossRef]
54. Ghai, R.R. *Interactions Between Primates and Parasites in a Wild Community*; McGill University (Canada): Montréal, QC, Canada, 2014.
55. Eke, S.S.; Omalu, I.C.J.; Ochaguba, J.E.; Urama, A.C.; Hassan, S.C.; Otuu, C.A.; Okafor, I.D. Prevalence of gastrointestinal parasites of sheep and goats slaughtered in Minna Modern Abattoir, Niger State, Nigeria. *J. Anim. Sci. Vet. Med.* **2019**, *4*, 65–70. [CrossRef]
56. Brunet, S. Analyse des Mécanismes D'action Antiparasitaire de Plantes Riches en Substances Polyphénoliques sur les Nématodes du Tube Digestif des Ruminants. Ph.D. Thesis, Paul Sabatier University, Toulouse, France, 2008.
57. Folstad, I.; Karter, A.J. Parasites, bright males, and the immunocompetence handicap. *Am. Nat.* **1992**, *139*, 603–622. [CrossRef]
58. Oppliger, A.; Giorgi, M.; Conelli, A.; Nembrini, M.; John-Alder, H. Effect of testosterone on immunocompetence, parasite load, and metabolism in the common wall lizard (*Podarcis muralis*). *Can. J. Zool.* **2004**, *82*, 1713–1719. [CrossRef]
59. Opeyemi, O.; Shittu, O.; Abdulganiyu, K.; Ashaolu, A.; Lawal, A.; Kadir, R. Helminth Infections of Captive Animals and Management Practices at the University of Ilorin Zoo, North-Central, Nigeria. *Niger. J. Parasitol.* **2022**, *43*, 214. [CrossRef]
60. Adrus, M.; Zainudin, R.; Ahamad, M.; Jayasilan, M.; Abdullah, M.T. Gastrointestinal parasites of zoonotic importance observed in the wild, urban, and captive populations of non-human primates in Malaysia. *J. Med. Primatol.* **2019**, *48*, 22–31. [CrossRef]
61. Zanzani, S.A.; Gazzonis, A.L.; Epis, S.; Manfredi, M.T. Study of the gastrointestinal parasitic fauna of captive non-human primates (*Macaca fascicularis*). *Parasitol. Res.* **2016**, *115*, 307–312. [CrossRef]

62. Berrilli, F.; Prisco, C.; Friedrich, K.G.; Di Cerbo, P.; Di Cave, D.; De Liberato, C. *Giardia duodenalis* assemblages and *Entamoeba* species infecting non-human primates in an Italian zoological garden: Zoonotic potential and management traits. *Parasites Vectors* **2011**, *4*, 307–312. [CrossRef]
63. Köster, P.C.; Dashti, A.; Bailo, B.; Muadica, A.S.; Maloney, J.G.; Santín, M.; Chicharro, C.; Migueláñez, S.; Nieto, F.J.; Cano-Terriza, D.; et al. Occurrence and genetic diversity of protist parasites in captive non-human primates, zookeepers, and free-living sympatric rats in the Córdoba Zoo Conservation Centre, Southern Spain. *Animals* **2021**, *11*, 700. [CrossRef]

Journal of
Zoological and Botanical Gardens

MDPI

Article

Parasitic Protozoa and Other Vector-Borne Pathogens in Captive Mammals from Brazil

Anisleidy Pérez Castillo [1,2], Nicolas Colácio [2], Pedro Henrique Cotrin Rodrigues [2], João Victor Oliveira Miranda [3], Paula Cristina Senra Lima [4], Rafael Otávio Cançado Motta [4], Herlandes Penha Tinoco [4], Carlyle Mendes Coelho [4] and Júlia Angélica Gonçalves da Silveira [2,*]

[1] Departamento de Parasitologia, Instituto de Ciências Biológicas, Universidade Federal de Minas Gerais, Belo Horizonte 31270–901, MG, Brazil; apeca81@gmail.com

[2] Laboratório de PROTOVET, Departamento de Medicina Veterinária Preventiva, Escola de Veterinária da Universidade Federal de Minas Gerais, Belo Horizonte 31270–901, MG, Brazil; nicocolacio@gmail.com (N.C.); pedro.cotrinr@gmail.com (P.H.C.R.)

[3] Laboratório de Biologia Integrativa, Departamento de Genética, Ecologia e Evolução, Instituto de Ciências Biológicas, Universidade Federal de Minas Gerais, Belo Horizonte 31270–901, MG, Brazil; jvmirandaoliveira76@gmail.com

[4] Fundação de Parques Municipais e Zoobotânica-FPMZB, Belo Horizonte 30210–090, MG, Brazil; paula.senra@gmail.com (P.C.S.L.); rafael.motta@pbh.gov.br (R.O.C.M.); herlandes@pbh.gov.br (H.P.T.); carlyle.m@pbh.gov.br (C.M.C.)

* Correspondence: jangelicags@gmail.com

Abstract: In captive environments, mammals are frequently exposed to various parasitic protozoa and other vector-borne pathogens that can impact both animal health and public health. Monitoring these pathogens is essential for animal welfare and zoonotic disease control. This study aimed to investigate the prevalence of parasitic protozoa and other vector-borne pathogens in captive mammals through molecular detection methods at the Belo Horizonte Zoo, Brazil. Between November 2021 and March 2023, whole blood samples were collected from 40 mammals. Molecular analyses identified piroplasms, *Leishmania* spp., granulocytic/platelet *Anaplasma/Ehrlichia* spp., monocytic *Ehrlichia* spp., *Bartonella* spp. and hemotropic *Mycoplasma* spp. with a 72.5% positivity rate. Piroplasms were found in 22.5% (two Pantanal cats, two gorillas, one white rhinoceros, one spider monkey, one jaguar, one tufted capuchin and one hippo) and *Leishmania* spp. in 12.9% (four maned wolves). Granulocytic/platelet *Anaplasma/Ehrlichia* spp. were found in 12.5% of the samples (one gorilla and four maned wolves), *Ehrlichia canis* in 2.5% of the animals (one maned wolf), *Bartonella* spp. in 42.5% (six howler monkeys, two maned wolves, one gorilla, one white rhino, one southern tamandua, one common woolly monkey, one tufted capuchin, one brown brocket deer, one agouti, one cougar and one hippo), hemotropic *Mycoplasma* spp. in 17.5% (one gorilla, one maned wolf, one white rhino, one howler monkey, two common woolly monkeys and one European fallow deer). Five Artiodactyla members tested negative for *A. marginale*. Coinfections occurred in 34.5% of the positive samples. Sequencing revealed that *Theileria* spp. and *Cytauxzoon* spp. are closely related to *Theileria bicornis* and *Cytauxzoon felis; Ehrlichia canis* and *Bartonella* spp. are closely related to *B. clarridgeiae* and *B. henselae;* and hemotropic *Mycoplasma* spp. are closely related to *Candidatus* Mycoplasma haemominutum. Our results showed a high occurrence of vector-borne pathogens in captive animals, including zoonotic species, which may pose a risk to animal and human public health.

Keywords: diagnosis; hemopathogens; captive environments; wildlife conservation; animal health; one health

1. Introduction

Diseases caused by vector-borne pathogens represent a significant challenge to the health and well-being of animals in captivity [1], with direct implications for the management of collections in zoos and conservation centers. These diseases can also affect the human population due to their zoonotic potential. The adverse effects of parasitic infections in zoo animals can vary [2], including the emergence of secondary deficiencies and other infections, reproductive impairment and the risk of mortality in cases of massive and dangerous parasitosis. Additionally, zoos are environments where wild animals have close contact with humans, considerably increasing the risk of transmission of parasitic zoonoses [3]. This poses a threat not only to the health of the animals themselves but also to the health of zoo staff and visitors.

The confined nature of these environments increases the susceptibility of animals to vector-borne infections due to population density, close interactions between species and environmental stress [4]. Additionally, the diversity of host species maintained in zoos provides a conducive environment for the transmission and maintenance of a wide range of vector-borne pathogens. The presence of these pathogens in zoo animals can result in a series of negative health impacts on the host, including anemia, weight loss, tissue damage, immunosuppression and, in severe cases, mortality [5].

Piroplasmids, such as *Babesia*, *Theileria* and *Cytauxzoon*, belong to the phylum Apicomplexa and are transmitted by parasitic ticks to mammals. These protozoa may be asymptomatic in wild animals but cause severe consequences and even death in domestic animals. These infections result in economic losses in livestock, trade restrictions in horses and negative health impacts on dogs and cats, as well as growing concern over cases of human babesiosis [6]. The genus *Leishmania*, member of the family Trypanosomatidae within the order Kinetoplastida, comprises protozoan pathogens transmitted by sandflies, such as *Lutzomyia* sp. The *Leishmania* protozoa infect both humans and a wide range of animals, contributing to the complexity and severity of zoonotic diseases and their impacts on public and veterinary health.

The Anaplasmataceae family, which includes the genera *Anaplasma*, *Ehrlichia*, *Neorickettsia* and *Wolbachia*, infects a variety of mammalian cells and can be transmitted by a wide range of vectors, such as ticks and trematodes. In Brazil, several species from this family have been detected in a wide range of wild animals, such as deer, wild canids, wild felids, coatis, rodents, peccaries, collared peccaries and opossums [7,8]. *Bartonella* species, belonging to the family Bartonellaceae, are transmitted by ectoparasites, such as fleas, lice and ticks, and can infect a variety of captive animals, including marine mammals such as dolphins as well as terrestrial mammals and their arthropod ectoparasites. The greater occurrence in some captive cohorts compared to free-ranging animals suggests that captivity may be a risk factor for *Bartonella* infection [9]. Additionally, hemotropic *Mycoplasma* species, classified within the class Mollicutes, are known to cause anemia in various mammalian species, transmitted by blood-feeding arthropods [10].

Vector-borne pathogens, including protozoa and bacteria, have been increasingly reported in captive mammals worldwide, with significant implications for both conservation efforts and animal welfare. For example, Protozoan agents have been associated with mortality and the development of clinical signs in captive cervids and canids from Canada and Brazil [11,12]. Bacterial pathogens have already been detected in captive animals around the world, Hemoplasmas have been detected in various captive animals, indicating that these pathogens can infect a wide range of species in captivity, which could potentially lead to health issues if not monitored [9,13–16].

Thus, the present work aimed to investigate the occurrence of vector-borne pathogens in captive mammals from the Zoo of the Belo Horizonte Zoo-Botanical Foundation, Minas Gerais, Brazil.

In this study, we report the high prevalence of vector-borne pathogens in various mammalian species at the Belo Horizonte Zoo by employing molecular techniques to detect and identify these pathogens. Our findings underscore the complexity of disease management in captive environments and the importance of early detection and integrated control measures to protect both animal and human health.

2. Materials and Methods

2.1. Ethical Statement

All procedures were conducted in accordance with relevant guidelines and regulations. The samples were collected after the signing of a partnership agreement between the Federal University of Minas Gerais (UFMG) and the Municipal Parks and Zoo-botanical Foundation (FPMZB) in strict compliance with applicable ethical guidelines.

2.2. Study Area

This study was conducted at the Belo Horizonte Municipal Parks and Zoo-Botanical Foundation (FPMZB), a 10.7-million-square-meter complex that houses more than 3500 individuals across 235 species, including 36 mammal species with 117 individuals. Notably, more than 40 species spanning reptiles, birds, fish, amphibians and mammals from all continents are threatened with extinction. Located in Belo Horizonte, Minas Gerais (−19.857889841, −44.0075113912) (Figure 1), the Zoo features a veterinary hospital and a rich history of biodiversity preservation. Established through Decree 16,684 of 31 August 2017, the FPMZB merged the former Municipal Parks Foundations and the Zoo-Botanical Foundation, playing a vital role in conserving local ecosystems and researching mammal species. Its parks and botanical gardens harbor more than a thousand species, representing Brazil's Cerrado and Atlantic Forest biomes [17].

2.3. Samples

Between November 2021 and March 2023, whole blood samples were collected from 40 captive animals of the class Mammalia at the Zoo. These samples were sent for analysis to the Veterinary Protozoology Laboratory—ProtoVet, as part of the service provided to FPMZB under the code Siex-UFMG 302557—"Diagnosis and control of hemoparasites in domestic and wild animals". The capture and containment of the animals were conducted by the Zoo's team of veterinarians following the management schedule of the mammal population. To ensure the representativeness of the sampling, at least one sample of each mammal species present in the zoo was collected.

2.4. DNA Extraction and PCR Amplification

DNA was extracted from 300 µL of whole blood using the "Wizard Genomic DNA Purification Kit" (Promega, Madison, WI, USA) according to the manufacturer's recommendations. After the extraction, the quality and quantity of the samples were estimated using a NanoDrop (Epoch Microplate Spectrophotometer, Biotek®, Winooski, Vermont, USA). The extracted DNA was stored at −20 °C until detection via PCR amplification. A fragment of mammalian glyceraldehyde-3-phosphate dehydrogenase (GAPDH) was used to verify the quality of the extracts [18].

Blood DNA samples from mammals that tested positive for the GAPDH gene were subjected to PCR assays for Piroplasmids, *Leishmania*, *Anaplasma*, *Ehrlichia*, *Bartonella* and hemotropic *Mycoplasma*. All agents, target genes, primer sequences, amplicon sizes (bp), thermal cycling conditions and references are shown in Table 1.

Table 1. Molecular assays used in the present study are categorized by the involved agent, type of PCR, target gene, primer sequences, amplicon size (bp), thermal cycling conditions and references.

Agents	Aim/Primer	Molecular Assay	Primers Sequences	Fragment Size (bp)	Thermal Cycling	Reference
***Anaplasma* sp.** (*A. phagocytophilum*, *A. bovis*, *A. platys*) (*16S rRNA* gene)	Screening 1° reaction gE3a gE10R	nPCR	5′-CACATGCAAGTCGAACGGATTATTC-3′ 5′-TTCCGTTAAGAAGGATCTAATCTCC′-3′	932	94 °C for 5 min 40 cycles: 94 °C for 30 s, 55 °C for 30 s and 72 °C for 1 min, 72 °C for 5 min	[19]
	2° reaction gE2 gE9f.		5′-GGCAGTATTAAAAGCAGCTCCAGG-3′ 5′-AACGGATTATTCTTTATAGCTTGCT-3′	546		
***Ehrlichia* sp.** (*E. chaffeensis*, *E. canis*) (*16S rRNA* gene)	Screening 1° reaction NS16SCH1F NS16SCH1R	nPCR	5′-ACGGACAATTGCTTATAGCCTT 5′-ACAACTTTTATGGATTAGCTAAAT	1195	94 °C for 5 min 30 cycles: 92 °C for 1 min, 54 °C for 1 min and 72 °C for 2 min, 72 °C for 8 min	[20]
	2° reaction NS16SCH2F NS16SCH2R		5′-GGGCACGTAGGTGGACTAG-3′ 5′-CCTGTTAGGAGGGATACGAC-3′	443		
Anaplasma phagocytophilum (*msp4* gene)	Characterization 1° reaction MSP4AP5 MSP4AP3	nPCR	5′-ATGAATTACAGAGAATTGCTTGTAGG-3′ 5′-TTAATTGAAAGCAAATCTTGCTCCTATG-3′	849	94 °C for 5 min 30 cycles: 92 °C for 1 min, 54 °C for 1 min and 72 °C for 2 min, 72 °C for 8 min	[21]
	2° reaction msp4F msp4R		5′-CTATTGGYGGNGCYAGAGT-3′ 5′-GTTCATCGAAAATTCCGTGGTA-3′	381		[22]
Anaplasma marginale*/** ***A. ovis (*msp4* gene)	Characterization 1° reaction MSP45 MSP43	nPCR	5′-GGGAGCTCCTATGAATTACAGAGAATTGTTTAC-3′ 5′-CCGGATCCTTAGCTGAACAGGAATCTTGC-3′	872	94 °C for 5 min 30 cycles: 92 °C for 1 min, 54 °C for 1 min and 72 °C for 2 min, 72 °C for 8 min	[23]
	2° reaction AnapF AnapR		5′-CGCCAGCAAACTTTTCCAAA-3′ 5′-ATATGGGGACACAGGCAAAT-3′	294		[14]
Anaplasma platys (*16S rRNA* gene)	Characterization 1° reaction 8-F 1448-R	nPCR	5′-AGTTTGATCATGGCTCAG-3′ 5′-CCATGGCGTGACGGGCAGTGT-3′	*	1° reaction 94 °C for 2 min, 40 cycles: 94 °C for 1 min, 45 °C for 1 min, 72 °C for 40 s, 72 °C for 4 min.	[24]
	2° reaction PLATYS-F EHR16S-R		5′-GATTTTTGTCGTAGCTTGCTATG-3′ 5′-TAGCACTCATCGTTTACAGC-3′	678	2° reaction 94 °C for 1 min, 40 cycles: 94 °C for 1 min, 53 °C for 30 s, 72 °C for 30 s, 72 °C for 4 min.	
Babesia*/*Theileira*/** ***Cytauxzoon (*18S rRNA* gene)	Screening 1° reaction RIB-19 RIB-20	nPCR	5′-CGGGATCCAACCTGGTTGATCCTGC-3′ 5′-CCGAATTCCTTGTTACGACTTCTC-3′	1700	94 °C for 5 min 30 cycles: 92 °C for 1 min, 54 °C for 1 min and 72 °C for 2 min, 72 °C for 8 min	[25]
	2° reaction BabRumF BabRumR		5′-ACCTCACCAGGTCCAGACAG-3′ 5′-GTACAAAGGGCAGGGACGTA-3′	430		[26]
Cytauxzoon felis	Characterization CytauxF CytauxR	cPCR	5′-CGAATCGCATTGCTTTATGCTCCAA 5′-TTGATACTCCGGAAAGAG	284	95 °C for 5 min, 40 cycles; 95 °C for 45 s, 59 °C for 45 s and 72 °C for 60 s, 72 °C for 5 min	[27]
***Leishmania* sp.** (*ITS1* gene)	Screening LITSR L5.8S	cPCR	5′-CTGGATCATTTTCCGATG-3′ 5′-TGATACCACTTATCGCACTT-3′	300–350	95 °C for 2 min, 37 cycles: 94 °C for 30s, 53 °C for 1 min, 72 °C for 1 min, 72 °C for 6 min.	[28]
***Bartonella* spp.** (16S-23S rRNA intergenic region)	Screening BartF BartR	cPCR	5′-CTCTTTCTTCAGATGATGATCC-3′ 5′-AACCAACTGAGCTACAAGCCCT-3′	145–232	95 °C for 2 min, 45 cycles: 95 °C for 1 min, 60 °C for 1 min, 72 °C for 30 s, 72 °C for 5 min	[29]
Hemotropic *Mycoplasma* sp. (*16S rRNA*)	Screening HBTF HBTR	cPCR	5′-ATACGGCCCATATTCCTACG-3′ 5′-TGCTCCACCACTTGTTCA-3′	618	94 °C for 10 min, 40 cycles: 95 °C for 30 s, 60 °C for 30 s, 72 °C for 30 s, 72 °C for 10 min	[30]

* the fragment size is unknown or cannot be determined.

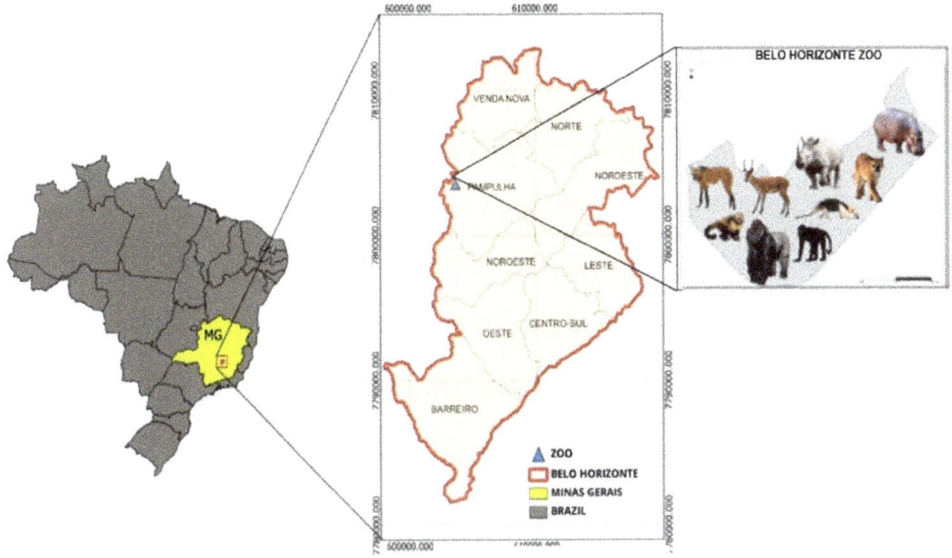

Figure 1. Capture site. Map of Minas Gerais state, southern Brazil, representing the Zoo region, where animals were sampled. Due to limitations in the software used to generate the figure, the numbers are not formatted with commas for thousands separation. Please disregard the format in the figure and refer to the correct formatting presented in the text (e.g., 7,780,000.000; 7,790,000.000; 7,800,000.000; 7,810,000.000; 600,000.000; 610,000.000).

For the PCRs, positive DNA controls were obtained from various sources. For tests of *Anaplasma phagocytophilum* and other granulocytic agents of the Anaplasmataceae family, DNA from IDE8 cell cultures infected with *A. phagocytophilum*, which was isolated from a German dog and kindly provided by Dr. Erich Zweygarth (Institut für Vergleichende Tropenmedizin und Parasitologie, Ludwig Maximilians Universität München), was used. DNA from *Anaplasma marginale* was extracted from 300 µL of whole blood collected from a calf experimentally infected with *A. marginale* (strain UFMG1) [31]. For monocytic ehrlichiosis tests, DNA from a dog infected with *Ehrlichia canis* (Jaboticabal strain) [32] was used. For the nPCR of piroplasmids, positive controls were obtained from a calf experimentally infected with *Babesia bovis* (strain BbovMG) and *Babesia bigemina* (strain BbigMG) [33]. DNA from reference *Leishmania* strains provided by the World Health Organization (WHO), specifically *Leishmania infantum* (MCAN/BR/2002/BH400), which was maintained in the cryobank of the *Leishmania* Biology Laboratory and kindly provided by Prof. Maria Norma Melo (ICB/UFMG/Brazil), was used as a reaction control. Additionally, DNA for hemoplasmas and *Bartonella* spp. was obtained from the blood samples of naturally infected cats in Belo Horizonte, Minas Gerais, Brazil, confirmed by sequencing [34]. Ultrapure sterile water (Life Technologies®, Carlsbad, CA, USA) was used as a negative control in all PCR assays.

PCR amplicons were separated by electrophoresis on 1%, 2% and 3% agarose gels (40 min, 100 V), stained with GelRedTM (Biotium, Hayward, CA, USA) and visualized under ultraviolet light.

2.5. Restriction Fragment Length Polymorphism (RFLP) Analysis—PCR for Leishmania spp.

To identify *Leishmania* spp., the *ITS1* amplicon was digested using the restriction enzyme HaeIII (PCR-RFLP). Five microliters of each amplified product were digested with the HaeIII restriction enzyme (Biolabs, Inc., UK) according to the manufacturer's instructions. The digested products were subjected to 5% polyacrylamide gel electrophoresis to verify the restriction patterns. The restriction patterns obtained were compared with those of WHO reference strains [28].

2.6. Sequencing

Sequencing was performed by ACTGene using an automatic sequencer (ABI 3730xl DNA Analyzer Applied Biosystems ™) with POP7 polymers and BigDye v3.1 using the same oligonucleotide primers used in the assays. For sequencing analysis, the raw data in the form of chromatograms generated by the automatic sequencer "Analyzer Applied Biosystems" were aligned, edited and analyzed using the BioEdit program, version 7.0.5.3 [35]. The identity of each sequence was confirmed by comparison with sequences available in GenBank using BLAST software (Basic Local Alignment Search Tool, http://blast.ncbi.nlm.nih.gov/ (accessed on 26 August 2024)) [36]. After comparing the identities, the analyzed sequences were classified according to the degree of similarity with data already deposited in GenBank. All nucleic acid sequences revealed in this study have been deposited in the GenBank database (PP844851, PP948683-PP948688, PP849127, PP849128, PP843077 and PP843078).

2.7. Phylogenetic Analysis

Multiple sequence alignments were carried out using the Multiple Alignment using Fast Fourier Transform (MAFTT) algorithm (MAFFT v7.310) [37], with previously published sequences in GenBank. Moreover, representative sample sequences were also downloaded for use as an outgroup in the phylogenetic analysis. Phylogenetic trees were constructed using the Maximum Likelihood (ML) method based on the results of alignment with MEGA11 [38–41] with suitable models selected with the smallest Bayesian information criterion (BIC) score. The reliability of the tree topology was tested using 1000 bootstrap replicates, as implemented in the program [42]. For piroplasmids, the Kimura 2-parameter method + G evolutionary model (K2 + G (5 categories, parameter = 0.1288)) (430 bp alignment) were used. For *Ehrlichia* spp., the Kimura 2-parameter model + gamma distribution (K2 + *G* (5 categories, parameter = 0.3607)) evolutionary model (630 bp) were used. For *Bartonella* spp., the Jukes–Cantor evolutionary model + gamma distribution + invariable sites (JC + G (5 categories, parameter = 1.0975)) + (I ([+*I*], 19.19% sites)) (165 bp alignment) were used, and for *Mycoplasma* spp., the general time reversible model with gamma distribution (GTR +G (5 categories, parameter = 0.3121)) (alignment of 600 bp).

2.8. Data Analysis

The data obtained were collected on an Excel spreadsheet, and descriptive frequency analysis was performed using Microsoft Excel Version 16.71 (Microsoft Corporation, Redmond, WA, USA).

3. Results

In this study, samples were collected from 40 individuals of the class Mammalia. Of these, 57.5% (23/40) were males, and 42.5% (17/40) were females. The animals under study formed a highly heterogeneous group due to the diversity of species found in the zoo, belonging to different orders: 45% (18/40) from the order Primates, 35% (14/40) from the order Carnivora, 12.5% (5/40) from the order Artiodactyla, 2.5% (1/40) from the order Perissodactyla, 2.5% (1/40) from the order Pilosa and 2.5% (1/40) from the order Rodentia. Details regarding the collected animal species can be found in Table 2.

Among the 40 DNA blood samples collected from mammals, all tested positive by GAPDH gene-based cPCR and were included in subsequent analyses.

Overall, 72.5% (29/40) of the samples tested positive for at least one parasite. Nine animals (9/40–22.5%) tested positive by nPCR for piroplasmids (two Pantanal cats (*Leopardus braccatus*), two gorillas (*Gorilla gorilla gorilla*), one white rhinoceros (*Ceratotherium simum*), one spider monkey (*Ateles* spp.), one jaguar (*Panthera onca*), one tufted capuchin (*Sapajus apella*) and one hippo (*Hippopotamus amphibius*). Sequencing was performed on all positive samples, but only two were successful. *Theileria* sp. was detected in a 54-year-old female white rhinoceros, showing 99.55% identity with *Theileria bicornis* found in a captive white rhinoceros (MF536661) from Australia. Additionally, a female jaguar was found to have a piroplasm with 99.77% identity to *Cytauxzoon felis*, which was previously identified in an ocelot (*Leopardus pardalis*) (GU903911) from Brazil. cPCR for *Leishmania* spp. was performed on 31 animals from the orders Primates and Carnivora, resulting in 12.9% positivity (four maned wolves (*Chrysocyon brachyurus*)), of which *Leishmania infantum* was detected in 50% (2/4) by PCR-RFLP.

Table 2. Identification of captive mammals sampled at Zoo of the Belo Horizonte Municipal Parks and Zoo-Botanical Foundation, Minas Gerais, between 2021 and 2023, regarding identification number, collection data, sex and molecular positivity for vector-borne pathogens. Results obtained in the PCR assays for *Anaplasma phagocytophilum*, *A. platys* and *Anaplasma marginale* are not shown since all animals tested negative.

ID	Collection Data	Sex	Species	Granulocytic/Platelet *Anaplasma*/*Ehrlichia* sp.	Monocytic *Ehrlichia* sp.	*Bartonella* sp.	Hemotropic *Mycoplasma* sp.	Piroplasmids	*Leishmania* spp.
AZ01	2021-11-04	M *	*Leopardus braccatus*	–	–	–	–	Positive	–
AZ02	2021-11-04	M	*Leopardus braccatus*	–	–	–	–	Positive	–
AZ04	2021-11-22	F **	*Gorilla gorilla gorilla*	–	–	Positive	Positive	Positive	–
AZ05	2021-11-22	F	*Gorilla gorilla gorilla*	Positive	–	–	–	Positive	–
AZ06	2021-12-07	F	*Puma concolor*	–	–	–	–	–	–
AZ07	2021-12-16	F	*Chrysocyon brachyurus*	–	–	Positive	–	–	Positive
AZ08	2022-02-08	M	*Galictis cuja*	–	–	–	–	–	–
AZ09	2022-02-08	M	*Subulo gouazoubira*	–	–	–	–	–	–
AZ11	2022-03-10	F	*Chrysocyon brachyurus*	Positive	–	–	Positive	–	Positive
AZ12	2022-03-10	F	*Chrysocyon brachyurus*	Positive	–	–	–	–	Positive
AZ13	2022-03-10	M	*Chrysocyon brachyurus*	Positive	–	–	–	–	Positive
AZ14	2022-03-16	M	*Chrysocyon brachyurus*	Positive	Positive	Positive	–	–	–
AZ15	2022-03-16	M	*Chrysocyon brachyurus*	–	–	–	–	–	–
AZ16	2022-05-16	M	*Dama dama*	–	–	–	–	–	–
AZ17	2022-05-16	M	*Gorilla gorilla gorilla*	–	–	–	–	–	–
AZ18	2022-05-27	M	*Chrysocyon brachyurus*	–	–	–	–	–	–
AZ19	2022-05-16	F	*Ceratotherium simum*	–	–	Positive	Positive	Positive	–
AZ20	2022-05-16	F	*Alouatta* sp.	–	–		Positive	–	–
AZ21	2022-05-27	F	*Alouatta* sp.	–	–		–	–	–
AZ22	2022-07-27	F	*Tamandua tetradactyla*	–	–	Positive	–	–	–
AZ25	2022-08-25	F	*Panthera leo*	–	–	–	–	–	–
AZ27	2022-07-01	F	*Ateles* sp.	–	–	–	–	Positive	–
AZ28	2022-10-25	F	*Lagothrix lagotricha*	–	–	Positive	Positive	–	–
AZ29	2022-10-25	M	*Sapajus apella*	–	–	Positive	–	–	–
AZ30	2022-11-15	M	*Subulo gouazoubira*	–	–	Positive	–	–	–
AZ31	2022-11-21	F	*Panthera onca*	–	–	–	–	–	–
AZ32	2022-11-24	M	*Lagothrix lagotricha*	–	–	–	–	–	–
AZ33	2022-11-24	F	*Lagothrix lagotricha*	–	–	–	Positive	–	–
AZ34	2022-11-24	M	*Sapajus apella*	–	–	–	–	–	–
AZ35	2022-11-24	F	*Lagothrix lagotricha*	–	–	–	–	–	–
AZ36	2023-01-24	M	*Dama dama*	–	–	–	Positive	–	–
AZ37	2023-01-24	M	*Dasyprocta* sp.	–	–	Positive	–	–	–
AZ38	2023-01-26	M	*Alouatta* sp.	–	–	Positive	–	–	–
AZ39	2023-01-26	F	*Alouatta* sp.	–	–	Positive	–	–	–
AZ41	2023-01-26	F	*Alouatta* sp.	–	–	Positive	–	–	–
AZ42	2023-01-26	M	*Alouatta* sp.	–	–	Positive	–	–	–
AZ43	2023-01-26	M	*Alouatta* sp.	–	–	Positive	–	–	–
AZ44	2023-01-25	F	*Puma concolor*	–	–	Positive	–	–	–
AZ45	2023-02-10	M	*Alouatta* sp.	–	–	Positive	–	–	–
AZ46	2022-11-15	F	*Hippopotamus amphibius*	–	–	Positive	–	Positive	–

* Male; ** Female; – Negative.

Phylogenetic analysis using the ML method based on the *18S rRNA* gene of piroplasmids (Figure 2) positioned the *Theileria* spp. sequences detected in white rhinoceroses from this study within the same clade as the *T. bicornis* sequences identified in white and black rhinoceroses from Australia and South Africa. Moreover, the *Cytauxzoon* spp. sequence detected in the jaguars in this study was closely related to *C. felis* sequences found in jaguars, ocelots and domestic cats from Brazil, South Africa and the United States.

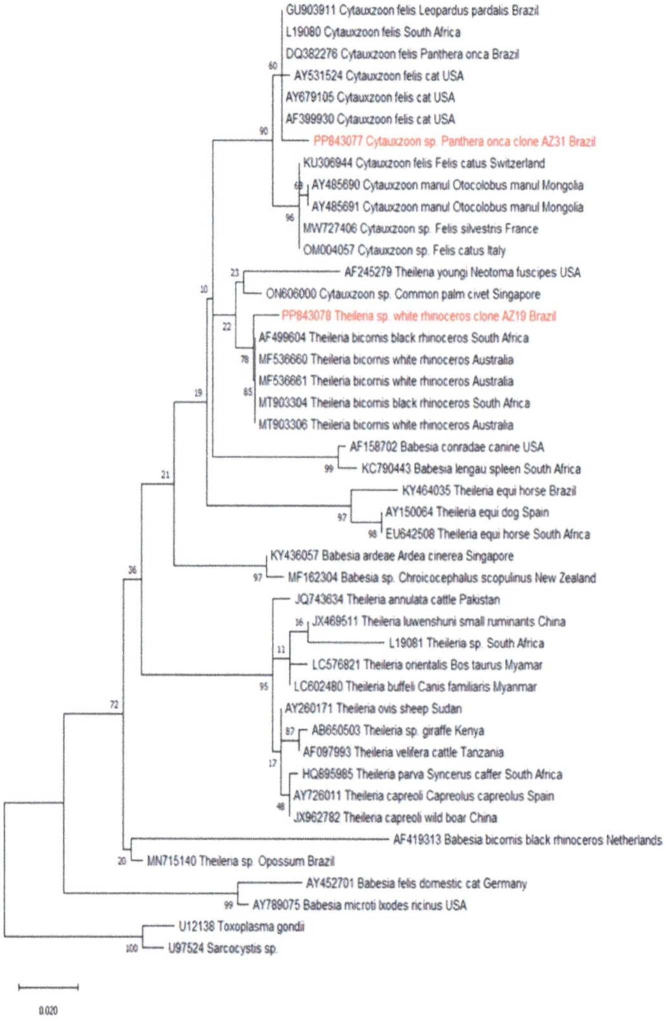

Figure 2. Phylogenetic tree based on an alignment of 430 bp of the piroplasmid *18S rRNA* gene involving 44 nucleotide sequences, using the ML method and K2 + G as an evolutionary model. The numbers at the branch nodes of the tree indicate bootstrap values from 1000 replications. The scale bar represents the evolutionary distance. The sequences detected in the present study are highlighted in red, with accession numbers provided in parentheses. *Toxoplasma gondii* and *Sarcocystis* spp. were used as outgroups.

In the nPCR analysis for the bacteria of the Anaplasmataceae family, granulocytic/platelet *Anaplasma/Ehrlichia* spp. were detected in 12.5% (5/40) of the animals, including one gorilla and four maned wolves. However, DNA amplification of *A. phagocytophilum* and *A. platys* was unsuccessful in these positive samples. Unfortunately, all samples positive for the *16S rRNA* gene by PCR for the *Anaplasma/Ehrlichia* spp. granulocytes/platelets yielded bands of weak intensity, which precluded sequencing efforts for these fragments. Additionally, 2.5% (1/40) of the animals tested positive for monocytic *Ehrlichia* spp. (one maned wolf), which was confirmed by sequencing similarity with *E. canis*. BLASTn analysis revealed 99.31% identity with the sequences of *E. canis* detected in *Canis lupus familiaris* (MK507008) sampled in Cuba.

ML analyses based on the *16S rRNA* gene of *Ehrlichia* sp. in this study detected a sequence that can be put into the same clade as previously identified *E. canis* sequences from domestic dogs in Brazil, Cuba, India and Turkey (Figure 3).

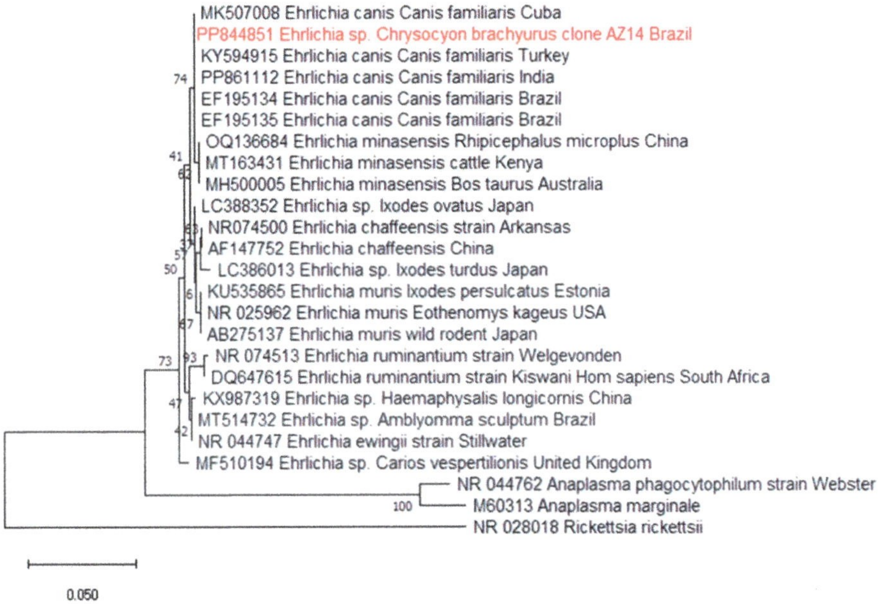

Figure 3. Phylogenetic tree based on an alignment of 630 bp of the *Ehrlichia* spp. *16S rRNA* gene involving 25 nucleotide sequences, using the maximum likelihood method and K2 + G as an evolutionary model. The numbers at the branch nodes of the tree indicate bootstrap values from 1000 replications. The scale bar represents the evolutionary distance. The sequences detected in the present study are highlighted in red, with accession numbers provided in parentheses. *A. marginale*, *A. phagocytophilum* and *Rickettsia rickettsii* were used as outgroups.

Bartonella spp. were detected in 42.5% (17/40) of the animals (six howler monkeys (*Alouatta* spp.), two maned wolves, one gorilla, one white rhino, one southern tamandua (*Tamandua tetradactyla*), one common woolly monkey (*Lagothrix lagotricha*), one tufted capuchin, one brown brocket deer (*Subulo gouazoubira*), one agouti (*Dasyprocta* spp.), one cougar (*Puma concolor*) and one hippo. Six samples were selected for sequencing due to the good intensity of the bands observed in the electrophoresis gel and the different sizes of the bands according to the primers used. The sequences obtained ranged from 152 bp to 345 bp, with four sequences showing 94.59% to 100% similarity to *Bartonella clarridgeiae* (the gorilla, tufted capuchin, brown brocket deer and southern tamandua) and two sequences showing 99% similarity to *Bartonella henselae* (Two howler monkeys).

Phylogenetic analysis using maximum likelihood, based on the internal transcribed spacer *16–23S* gene, grouped *Bartonella* spp. sequences detected in the gorilla, southern tamandua, capuchin monkey and brown brocket deer from this study into the same clade as *B. clarridgeiae* sequences from cats and fleas. However, *Bartonella* spp. sequences detected in two howler monkeys from this study clustered with *B. henselae* previously identified in cats, rodents, elephants, Fea's muntjacs and humans (Figure 4).

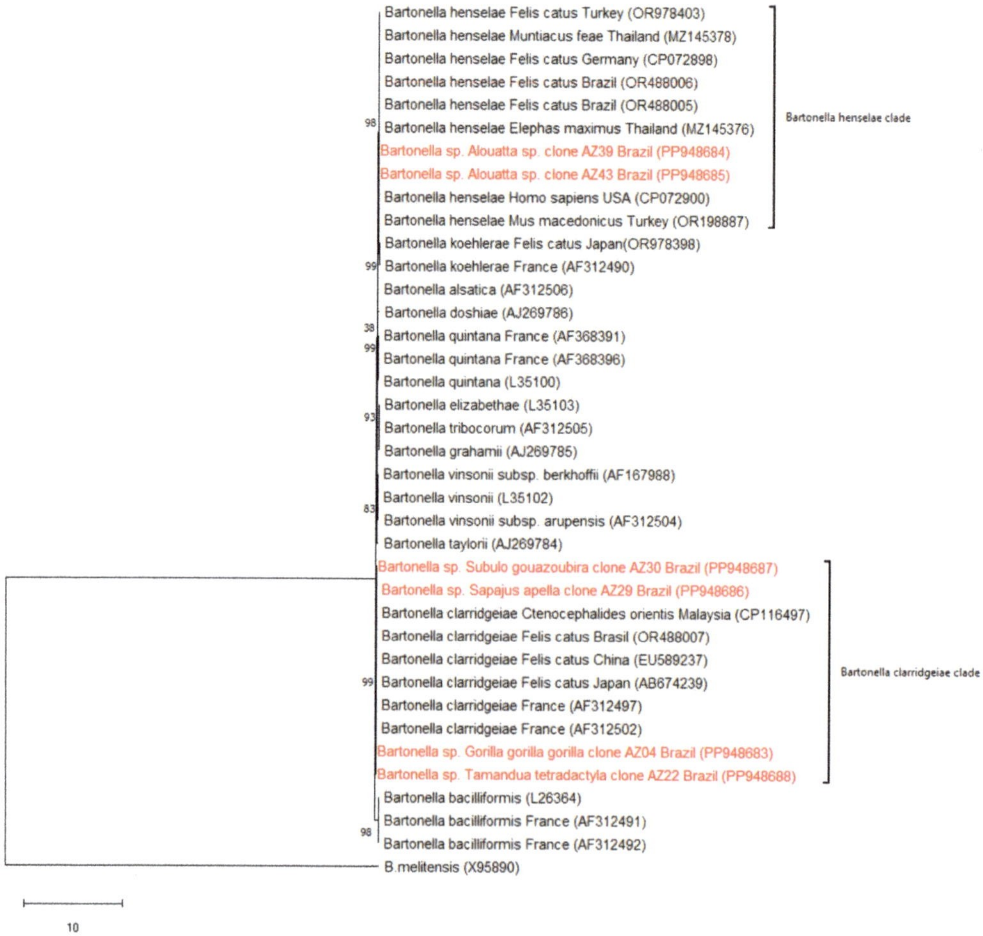

Figure 4. Phylogenetic tree based on an alignment of (~165 bp) bp of the *Bartonella* spp. 16S-23S rRNA intergenic region gene involving 38 nucleotide sequences, using the ML method and JC + G + I as an evolutionary model. The numbers at the branch nodes of the tree indicate bootstrap values from 1000 replications. The scale bar represents the evolutionary distance. The sequences detected in the present study are highlighted in red/underline, with accession numbers provided in parentheses. *Brucella melitensis* was used as an outgroup.

Hemotropics *Mycoplasma* spp. were detected in 17.5% (7/40) of the animals, including one gorilla, one maned wolf, one white rhino, one howler monkey, two common woolly monkeys and one European fallow deer (*Dama dama*). Two cases were confirmed by sequencing as hemotropic *Mycoplasma* sp.: one in the howler monkeys, showing 99.84% identity with "*Candidatus* Mycoplasma haemominutum" detected in a cat (KR905451) sampled in Italy, and another in a woolly monkey, showing 99.11% identity with a *Mycoplasma* sp. detected in a howler monkey (MH734376) sampled in Brazil.

Finally, ML analyses of hemotropic *Mycoplasma* spp. grouped the sequences detected in this study into a large clade, the "*Mycoplasma suis* group," alongside sequences previously detected in various hosts from several countries, with 100% clade support (Figure 5). The sequence detected in the howler monkey was closely related to that of Ca. Mycoplasma haemominutum, which had been found in domestic cats from Brazil and Italy.

Meanwhile, the sequence detected in a woolly monkey was closely related to hemotropic *Mycoplasma* spp. found in howler monkeys from Brazil.

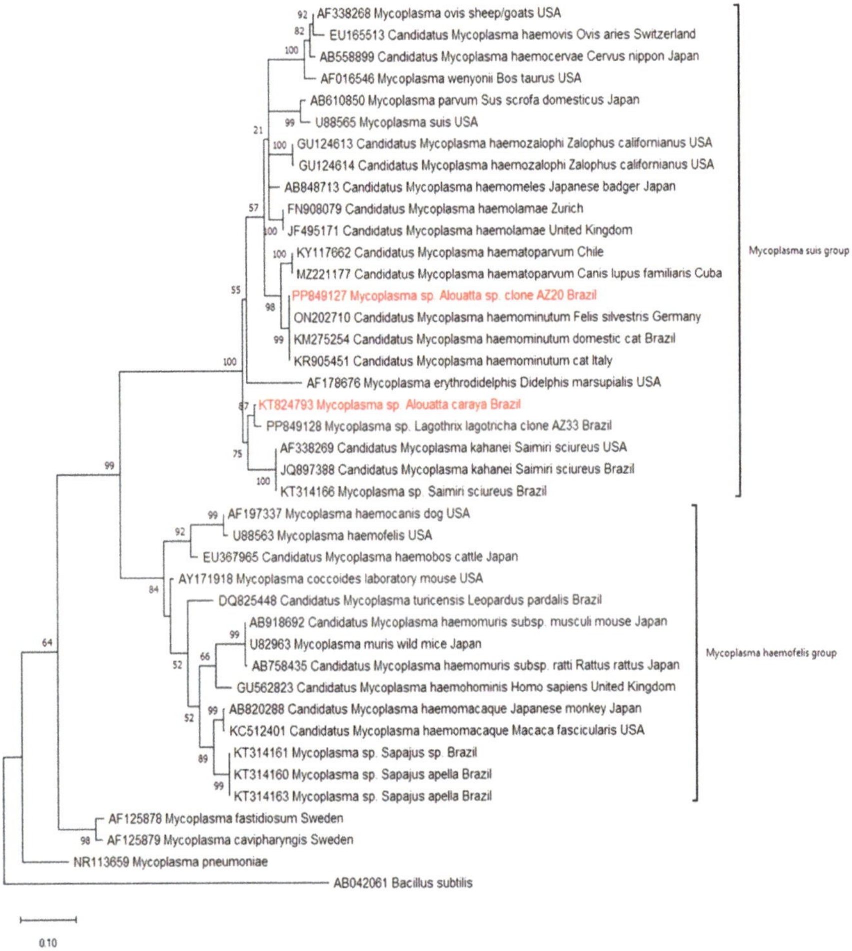

Figure 5. Phylogenetic tree based on an alignment of 600 bp of *Mycoplasma* spp. *16S rRNA* gene involving 41 nucleotide sequences, using the ML method and GTR + G as an evolutionary model. The numbers at the branch nodes of the tree indicate bootstrap values from 1000 replications. The scale bar represents the evolutionary distance. The sequences detected in the present study are highlighted in red, with accession numbers provided in parentheses. *Mycoplasma pneumoniae* and *Bacillus subtilis* were used as outgroups.

Coinfections Detected

Of the 29 positive animals, 10 (34.5%) were coinfected (five maned wolves, two gorillas, one white rhinoceros, one woolly monkey and one hippo). Six animals were coinfected with two pathogens: one gorilla was coinfected with piroplasmids and granulocytic/platelet *Anaplasma/Ehrlichia* spp., a single maned wolf was coinfected with *Leishmania* spp. and *Bartonella* spp., two maned wolves were coinfected with granulocytic/platelet *Anaplasma/Ehrlichia* spp. and *Leishmania* spp., a single woolly monkey was coinfected with hemotropic *Mycoplasma* spp. and *Bartonella* spp. and one hippopotamus was coinfected with piroplasmids and *Bartonella* spp. Four animals were coinfected with three pathogens: one gorilla and one white rhino were coinfected with piroplasmids, hemotropic *Mycoplasma* spp. and *Bartonella* spp.; one maned wolf was coinfected with granulocytic/platelet *Anaplasma/Ehrlichia* spp., hemotropic *Mycoplasma* spp. and *Leishmania* spp.; and one maned wolf was coinfected with *Ehrlichia canis*, granulocytic/platelet *Anaplasma/Ehrlichia* spp. and *Bartonella* spp. (Table 3).

Table 3. Coinfections in captive mammals sampled at the Zoo of the Belo Horizonte Municipal Parks and Zoo-Botanical Foundation, Minas Gerais, Brazil between 2021 and 2023.

ID	Species	Granulocytic/Platelet *Anaplasma/Ehrlichia* sp.	Monocytic *Ehrlichia* sp.	*Bartonella* sp.	Hemotropic *Mycoplasma* sp.	Piroplasmids	*Leishmania* spp.	Total Number of Pathogens Coinfected
AZ04	*Gorilla gorilla gorilla*	−	−	+	+	+	−	3
AZ05	*Gorilla gorilla gorilla*	+	−	−	−	+	−	2
AZ07	*Chrysocyon brachyurus*	−	−	+	−	−	+	2
AZ11	*Chrysocyon brachyurus*	+	−	−	+	−	+	3
AZ12	*Chrysocyon brachyurus*	+	−	−	−	−	+	2
AZ13	*Chrysocyon brachyurus*	+	−	−	−	−	+	2
AZ14	*Chrysocyon brachyurus*	+	+	+	−	−	−	3
AZ19	*Ceratotherium simum*	−	−	+	+	+	−	3
AZ28	*Lagothrix lagotricha*	−	−	+	+	−	−	2
AZ46	*Hippopotamus amphibius*	−	−	+	−	+	−	2

+ Positive, − Negative.

4. Discussion

Vector-borne diseases involving various infectious agents, hosts and vectors are highly important to human and animal health. Disease surveillance and control in zoological environments are crucial for preventing outbreaks and protecting both the species on display and nearby human populations [43]. The study results demonstrated a high prevalence of vector-borne pathogens in wild mammals at the Belo Horizonte Zoo-Botanical Foundation, with 72.5% of the tested samples showing positivity for at least one vector-borne pathogen.

It is worth noting that Belo Horizonte is considered an endemic region for most of the agents studied, which corroborates the high occurrence described here. Since the first report of *E. canis* in the city, the pathogen has spread throughout the territory and is being found in various regions of Brazil [44]. Additionally, *A. platys* is endemic in Brazil, with the prevalence varying according to the geographic region, target population and diagnostic methods used [45]. A study conducted in Minas Gerais with Belo Horizonte as the coverage area reported a prevalence of 4.1% and 17.2% of *A. platys* in dry and rainy periods, respectively, both in rural areas, while a lower prevalence was observed in urban areas (5%). Regional and climatic variations directly affect vector population density, providing specific conditions that influence the dissemination of the agent [33].

Canine visceral leishmaniasis (CVL) is a significant public health issue in Belo Horizonte, Brazil, with domestic dogs being the main reservoir host for the causative parasite *Leishmania infantum* [46,47]. Studies have shown that a large portion, estimated at 40–60%, of seropositive dogs are asymptomatic carriers. These asymptomatic dogs can still harbor the parasite, especially on their skin, and contribute to the maintenance of the disease transmission cycle [48]. Furthermore, research in Belo Horizonte has identified a high percentage of infected dogs that were asymptomatic and PCR positive but seronegative, posing challenges for health authorities in implementing effective control measures [49]. The prevalence of CVL in the city's neighborhoods can vary widely, from 0 to 166.7 cases per 1000 dogs [47]. Spatial and temporal analyses have been used to identify priority areas for targeted surveillance and control efforts [46]. Human visceral leishmaniasis (HVL) cases in Belo Horizonte have also been found to be associated with canine seroprevalence and the human-to-dog ratio, underscoring the importance of addressing the canine reservoir in reducing disease transmission to humans [49].

Despite strict regulations prohibiting the entry of domestic animals into the zoo, stray dogs and cats often infiltrate the area, potentially introducing ectoparasites and pathogens to captive animals [50]. This uncontrolled access increases the risk of transmission of vector-borne diseases, underscoring the importance of stringent biosecurity measures and continuous, comprehensive monitoring of both zoo and stray animals for the health of the captive population and potential public health risks.

For piroplasmids, 22.5% (9/40) of the tested animals were positive, with confirmed detections of *Theileria* sp. related to *T. bicornis* and *Cytauxzoon* sp. related to *C. felis*, indicating a diversity of piroplasma species infecting different hosts. The infection rates in our study align with the literature, which reports values ranging from 5.2% to 96.7% for piroplasms, including *Babesia* spp., *Theileria* spp. and *Cytauxzoon* spp., in wild mammals [51]. For instance, ref. [51] reported a 5.5% prevalence of piroplasms in free-ranging mammals of the Superorder Xenarthra from four Brazilian states, whereas another study in Brazil reported a prevalence of 26.3% (40/152) in wild mammals [52,53]. Ref. [53] reported a 96.7% prevalence in free-ranging jaguars [54]. These differences are possibly due to variations in host species, geographic location and environmental factors [55].

Theileria bicornis has been reported in black and white rhinoceroses in Africa, with implications for their health and conservation. *T. bicornis* infections can be fatal in rhinoceroses, particularly when combined with other stressors such as injury, pregnancy or translocation [56], while *C. felis* is known to infect felids such as jaguars in Brazil [57]. The available evidence strongly suggests that jaguars can act as reservoirs for *C. felis*, harboring and potentially transmitting the parasite to other susceptible hosts. Moreover, *C. felis* infections appear to be capable of causing fatal disease in jaguars, similar to the severe outcomes observed in domestic cats [58,59]. Despite its predominance in wild felids, cases of infection in domestic felids are increasing in Brazil [59].

In this study, we detected sequences of *Theileria* spp. in a 54-year-old female white rhinoceros originating from Africa, housed at the Belo Horizonte Zoo and donated from Germany. To date, there has not been a description of *T. bicornis* in this animal species in Brazil, suggesting that the animal arrived at the zoo with a pre-existing infection, although there are recent reports of a piroplasm with 91.8 to 93.6% identity with *T. bicornis* in bats [60]. Phylogenetic analysis grouped the sequences detected in our study in the same clade as sequences of *T. bicornis* identified in white and black rhinoceroses from Australia and South Africa [56,61]. The phylogenetic similarity between these sequences may indicate a common origin of these pathogens or transmission via vectors sharing a common habitat across these regions [62,63].

Furthermore, *Cytauxzoon* spp. sequences detected in jaguars were closely related to the sequences of *C. felis* found in jaguars, domestic cats and other wild felids from Brazil, South Africa and the United States. This finding supports the global distribution of this protozoan and indicates its ability to infect a variety of feline species across different geographic regions [54,58]. Although the pathogenicity of *Cytauxzoon* spp. genotypes found in Brazil is unknown, free-ranging jaguars likely play a crucial role in the survival of *C. felis* in the wild [54]. Therefore, continuous surveillance and monitoring of the spread of *C. felis* are important, especially in environments where multiple feline species coexist, such as in zoos [58].

The presence of *Leishmania* spp. in maned wolves is a significant finding, highlighting the zoonotic nature of this parasite [64]. Leishmaniasis, particularly visceral leishmaniasis, is a serious disease in both humans and animals. The presence of *Leishmania* spp. in maned wolves is consistent with findings in other wild canids, underscoring the epidemiology of this disease and the need for severe vector control techniques and continuous health monitoring to prevent outbreaks [65]. A study by Silveira et al. (2016) [63] reported a case of *Rangelia vitalii* in a free-ranging maned wolf, which also tested positive for *Leishmania* sp., among other pathogens [64]. Similarly, a study by Mol et al. (2015) [64]

demonstrated the transmissibility of *L. infantum* from maned wolves to the invertebrate vector *Lutzomyia longipalpis*, further emphasizing the role of wildlife in the epidemiology of leishmaniasis [65].

The detection of granulocytic/platelet *Anaplasma/Ehrlichia* spp. in 12.5% of the studied animals is consistent with previous reports of infections in wild and captive animals. Studies conducted by Calchi et al. (2020) and Pereira et al. (2016) [7,66] reported similar prevalences in wild animals in Brazil (Xenarthra mammals) and Portugal (cervids and wild boars), with prevalence rates of 27.57% and 24.82%, respectively. The unsuccessful amplification of DNA from *A. phagocytophilum* and *A. platys* might indicate the presence of *Anaplasma* species or genetic variants not detectable by the primers used, limited PCR sensitivity or a low parasitic load in the samples [67]. Previous studies have documented similar challenges in the molecular detection of *Anaplasma* species, where primer specificity and parasitic load significantly influence the success of amplification [68]. Although the DNA of *A. marginale* in members of the Artiodactyla order was not amplified in this study, this pathogen has already been detected in South American deer from Brazil [14].

The identification of *E. canis* in the maned wolf supports the hypothesis that stray dogs may act as carriers of arthropod-borne pathogens to captive wild animals. Serological and molecular studies have detected the antibodies and DNA of *E. canis* in free-ranging maned wolves in Brazil, indicating serious health risks due to the severe nature of ehrlichiosis [69]. Despite being collected from a maned wolf, a wild carnivore native to South America, the results showed a close phylogenetic relationship with strains found in domestic dogs (*C. familiaris*) from Brazil, Cuba, India and Turkey, underscoring their global distribution and the importance of disease surveillance and control. From a taxonomic perspective, the close phylogenetic relationship suggested that the sample may belong to the same species or a species closely related to *E. canis* [70].

This study detected *Bartonella* spp. in 42.5% (17/40) of the animals, including a variety of species. This finding aligns with previous research indicating a significant incidence of *Bartonella* in various mammalian species, both in wild and captive habitats [71,72]. The prevalence in this study is consistent with the findings of Rao et al. (2021) [73], who reported significant proportions of *Bartonella* (38.61%) among small mammals in the Qaidam Basin, western China. A study in Italian Nature Reserve Parks detected *Bartonella* infection in 97 red foxes, eight European badgers, six Eurasian wolves, six European hedgehogs, three beech martens and two deer. The prevalence was 9.84%, with zoonotic species detected in wolves (83.3%), hedgehogs (33.33%) and foxes (4.12%) [73]. *Bartonella rochalimae* was the most common species found in foxes and wolves. Zoonotic species were significantly more frequent in Eurasian wolves ($p < 0.0001$), indicating that they may be reservoirs for infection in humans and domestic animals [74]. In Brazil, several studies have investigated the occurrence of *Bartonella* in wild and domestic animals, revealing a wide diversity of genotypes and potential zoonotic risks [72,75].

ITS phylogenetic analysis grouped *B. clarridgeiae* sequences from cats and fleas with our sequences from gorillas, southern tamanduas, capuchin monkeys and brocket deer. Conversely, *B. henselae* sequences from cats, deer, rodents and humans clustered with our sequences from howler monkeys. These findings expand the known host range for *B. clarridgeiae* and *B. henselae*, suggesting potential underestimation of their prevalence in captive animals [29]. Despite the small sequences used for constructing the phylogeny, it was possible to distinguish between the species. The *ITS* gene, a noncoding region of ribosomal DNA, varies between *Bartonella* species, enabling their molecular differentiation [76].

Our study revealed a 17.5% prevalence of hemotropic *Mycoplasma* spp. among tested animals infected with multiple species. These bacteria have been detected in numerous nonhuman primates and captive mammals globally [77,78]. In Brazil, several investigations have focused on hemoplasma infections across diverse animal populations. For instance, captive *Allouata* in São Paulo's zoos exhibited a 26.47% positivity rate for hemoplasmas [79], while captive wild carnivores showed high prevalence rates, such as 45.5% in wild felids and 83.3% in wild canids [80]. Moreover, studies involving non-human primates, both captive and free-ranging, reported an infection rate of 25%, with notable prevalence among black howler monkeys (64.3%) and black-horned capuchins (4.2%) [77]. In the southern Pantanal region, *Mycoplasma* spp. infections were detected in various wild mammals, including *Nasua nasua, Cerdocyon thous* and *Leopardus pardalis,* indicating exposure to multiple hemoplasma species [10].

Moreover, our study identified hemotropic *Mycoplasma* spp. in *D. dama*, consistent with previous findings in cervids by André et al. (2020) [80], who reported these bacteria in 40% of free-ranging deer in Brazil, broadening our understanding of hemoplasma distribution in deer populations [81]. These results underscore the importance of further research to elucidate the ecology of these pathogens in captive populations, aligning with the One Health concept, given reports of infection in immunocompromised humans.

Phylogenetic analysis revealed that the sequences identified in this study for hemotropic Mycoplasma species are phylogenetically closely related to those of the *M. suis* group. Previous research has demonstrated that hemotropic *Mycoplasma* species form specific evolutionary groups based on their hosts. The placement of the sequences detected in non-human primates within the "*M. suis* group" clade is consistent with this pattern, suggesting a common ancestor between the species that infect NHPs and other animals, reflecting possible specific adaptations to their hosts [82]. Although "Ca. M. haemominutum" was detected by *16S rRNA* cPCR in the howler monkeys, presumably, this cPCR assay revealed a sequence of the novel *Mycoplasma* sp. because several regions of homology may be found with "Ca. M. haemominutum" sequences [77]. The *16S rRNA* gene can show 99% homology with sequences of other pathogenic *Mycoplasma* species from different hosts, suggesting a close relationship between the species. However, when other genes are used for molecular characterization, little identity is observed, indicating that they are different species. Therefore, it is necessary to characterize isolates using multiple genes to investigate genetic diversity and accurately determine the species [4,83].

The tree showed that the sequence detected in a captative woolly monkey in the Belo Horizonte Zoo, Minas Gerais, clustered within a clade with hemotropic *Mycoplasma* spp. detected in howler monkeys from southern Brazil. This suggests a possible relationship between the pathogens in different species of Neotropical primates and indicates that these hemoplasmas occur in close geographic locations [77,79]. Furthermore, phylogenetic trees have shown an evident separation of the hemoplasma detected in the woolly monkey from "Ca. M. kahanei" was found in squirrel monkeys in Brazil, and hemotropic *Mycoplasma* spp. were detected in *S. apella* in Brazil, indicating the presence of different species, an event that has been previously reported [77,79].

Our study revealed a high prevalence (34.7%) of coinfections by vector-borne pathogens in mammals at the Belo Horizonte Zoo, Brazil. Perles et al. (2023) [4] reported that only a small percentage (8.1%) of the studied animals tested negative for the evaluated agents, with bacterial coinfections being the most common (18.3%), particularly involving *Mycoplasma* spp. and *Bartonella* spp. Comparing it with our findings, we see those coinfections with bacterial pathogens, especially *Mycoplasma* spp. and *Bartonella* spp., are prevalent in both studies. Although vector-borne infections often remain subclinical in wild animals, they may act as opportunistic pathogens in immunocompromised animals under stressful conditions or in the presence of coinfections [4].

For example, hemotropic *Mycoplasma* species can cause severe anemia in cats with concurrent FeLV infection or immunosuppression [4,84]. Coinfections can complicate diagnosis and treatment and worsen prognosis in animals. In a study on cats, 22 out of 624 (4%) tested positive for more than one vector-borne pathogen, with nine cats infected with *Hepatozoon* spp. having antibodies against *Leishmania* spp., *Rickettsia* spp. and *Ehrlichia* spp. [84]. Additionally, the authors of [63] reported a fatal case of parasite coinfection in a threatened maned wolf from Minas Gerais, Brazil [64].

The detection of possible reservoirs of zoonotic pathogens in wildlife is crucial to public health because it allows us to anticipate and prevent the spread of diseases between animals and humans, in line with the concept of *One Health*. In a study conducted in

Switzerland, antibodies against SARS-COV-2 were found in red foxes, Eurasian lynxes and wild cats, indicating that these animals were exposed to the virus, but without signs of active infection [85]. Similarly, our research at the Belo Horizonte Zoo found a significant incidence of zoonotic infections such as *Bartonella* spp., *Mycoplasma* spp. and *Leishmania* spp. in captive mammals, reinforcing the potential risk of disease transmission between animals and humans. These results emphasize the importance of continually monitoring the circulation of zoonotic pathogens in different animal environments to preserve public health and prevent future pandemics.

Our study demonstrated that wild and exotic animals kept in zoos are also at risk of infection with several species of vector-borne pathogens. Our findings showed the high prevalence of these agents that infect and coinfect mammals at the Belo Horizonte Zoo in Brazil, reinforcing the urgent need to improve control measures, including vector control strategies and routine parasitological examinations, to protect animal health and prevent the spread of diseases, while advancing our understanding of parasite–host dynamics in animals, benefiting both public health and wildlife conservation.

Limitations of the present study should be considered. A notable limitation of the present study is that the phylogeny was based on a small fragment (165 to 630 bp) of a conserved gene, which precluded additional phylogenetic inferences to confirm species identification, potentially limiting the resolution of our analyses. Unfortunately, it was not possible to sequence more samples or amplify other gene targets to evaluate the genetic diversity between the vector-borne pathogens. Additionally, short-read sequencing platforms targeting regions of *16S rRNA* or *18S rRNA* may not achieve complete taxonomic resolutions consistently obtained through whole-genome sequencing [84]. Future studies should prioritize larger sample sizes and more detailed genetic analyses, including the use of other genes, to improve our understanding of phylogenetic relationships and species differentiation [86].

Although this study identified several pathogens in captive mammals in Brazil, it was not possible to evaluate all pathogens of interest, such as *Toxoplasma gondii*. Detection of *T. gondii* requires specific methods that were not implemented in this study. We recognize the importance of *T. gondii* in wildlife parasitology and public health studies due to its zoonotic potential and significant impact on domestic and wild animals [87,88]. Future research should include the evaluation of this and other relevant pathogens to provide a more comprehensive understanding of the epidemiology of parasitic diseases in captive environments.

Author Contributions: A.P.C.: Conceptualization, investigation, methodology, analysis, formal analysis, data curation, writing—original draft, writing, reviewing and editing. N.C.: Methodology, analysis, writing—review and editing. P.H.C.R.: Methodology, analysis, writing—review and editing. J.V.O.M.: Resources, writing the original draft, writing the review and editing. P.C.S.L.: Methodology. R.O.C.M.: Methodology. H.P.T.: Methodology. C.M.C.: Resources, investigation, supervision and funding acquisition. J.A.G.d.S.: Conceptualization, resources, investigation, supervision, funding acquisition, review and editing. All authors have read and agreed to the published version of the manuscript.

Funding: This research was funded by the Coordenação de Aperfeiçoamento de Pessoal de Nível Superior (CAPES), Fundação de Amparo à Pesquisa do Estado de Minas Gerais-FAPEMIG (APQ-00708–21 and APQ-02531–24), Conselho Nacional de Desenvolvimento Científico e Tecnológico-CNPq and Universidade Federal de Minas Gerais.

Institutional Review Board Statement: All procedures were conducted in accordance with relevant guidelines and regulations. The samples were collected after the signing of a partnership agreement between the Federal University of Minas Gerais (UFMG) and the Municipal Parks and Zoo-botanical Foundation (FPMZB) in strict compliance with applicable ethical guidelines. Samples were sent for analysis to the Veterinary Protozoology Laboratory—ProtoVet, as part of the service provided to FPMZB under the code Siex-UFMG 302557—"Diagnosis and control of hemoparasites in domestic and wild animals".

Data Availability Statement: The data presented in this study are available on request from the corresponding author.

Acknowledgments: Special thanks are extended to all employees of the laboratory PROTOVET, Residents in Public Health with emphasis on human and wildlife health interfaces, and to the mammal management team at the Municipal Parks and Zoo-Botanical Foundation—FPMZB—in Belo Horizonte.

Conflicts of Interest: The authors declare no conflicts of interest.

References

1. Li, H.; Chen, Y.; Machalaba, C.C.; Tang, H.; Chmura, A.A.; Fielder, M.D.; Daszak, P. Wild animal and zoonotic disease risk management and regulation in China: Examining gaps and One Health opportunities in scope, mandates, and monitoring systems. *One Health* **2021**, *13*, 100301. [CrossRef] [PubMed]
2. Ferdous, S.; Chowdhury, J.; Hasan, T.; Dutta, P.; Rahman, M.; Hassan, M.M.; Faruque, R.; Alim, M.A. Prevalence of gastrointestinal parasitic infections in wild mammals of a safari park and a zoo in Bangladesh. *Vet. Med. Sci.* **2023**, *9*, 1385–1394. [CrossRef] [PubMed]
3. Panayotova-Pencheva, M.S. Parasites in captive animals: A review of studies in some European zoos. *Der Zool. Gart.* **2013**, *82*, 60–71. [CrossRef]
4. Perles, L.; Moraes, M.F.; da Silva, M.X.; Vieira, R.F.C.; Machado, R.Z.; Hoppe, E.G.L.; André, M.R. Co-infection by multiple vector-borne agents in wild ring-tailed coatis (*Nasua nasua*) from Iguaçu National Park, southern Brazil. *Sci. Rep.* **2023**, *13*, 1828. [CrossRef]
5. Iatta, R.; Natale, A.; Ravagnan, S.; Mendoza-Roldan, J.; Zatelli, A.; Cavalera, M.A.; Nachum-Biala, Y.; Baneth, G.; Otranto, D. Zoonotic and vector-borne pathogens in tigers from a wildlife safari park, Italy. *Int. J. Parasitol. Parasites Wildl.* **2020**, *12*, 250–325. [CrossRef]
6. Almazán, C.; Scimeca, R.C.; Reichard, M.V.; Mosqueda, J. Babesiosis and Theileriosis in North America. *Pathogens* **2022**, *11*, 168. [CrossRef]
7. Calchi, A.C.; Vultão, J.G.; Alves, M.H.; Yogui, D.R.; Desbiez, A.L.J.; De Santi, M.; de Souza Santana, M.; Da Silva, T.M.V.; Werther, K.; Teixeira, M.M.G.; et al. *Ehrlichia* spp. and *Anaplasma* spp. in Xenarthra mammals from Brazil, with evidence of novel '*Candidatus anaplasma* spp.'. *Sci. Rep.* **2020**, *10*, 12615. [CrossRef]
8. Da Silveira, J.A.G.; Moreira, S.M.; Nascimento, A.F.D.; de Oliveira, M.M.; dos Santos, H.A.; Estevam, L.G.T.d.M.; Pereira, C.R.; Oliveira, A.G.G.; D'elia, M.L.; Araujo, A.d.C.; et al. Preparing Collared Peccary (*Pecari tajacu* Linnaeus, 1758) for Reintroduction into the Wild: A Screening for Parasites and Hemopathogens of a Captive Population. *Pathogens* **2024**, *13*, 47. [CrossRef]
9. Harms, C.A.; Maggi, R.G.; Breitschwerdt, E.B.; Clemons-Chevis, C.L.; Solangi, M.; Rotstein, D.S.; Fair, P.A.; Hansen, L.J.; Hohn, A.A.; Lovewell, G.N.; et al. Bartonella species detection in captive, stranded and free-ranging cetaceans. *Vet. Res.* **2008**, *39*, 59. [CrossRef]
10. De Sousa, K.C.M.; Herrera, H.M.; Secato, C.T.; Oliveira, A.D.V.; Santos, F.M.; Rocha, F.L.; Barreto, W.T.G.; Macedo, G.C.; Pinto, P.C.E.d.A.; Machado, R.Z.; et al. Occurrence and molecular characterization of hemoplasmas in domestic dogs and wild mammals in a Brazilian wetland. *Acta Trop.* **2017**, *171*, 172–181. [CrossRef]
11. Mathieu, A.; Pastor, A.R.; Berkvens, C.N.; Gara-Boivin, C.; Hébert, M.; Léveillé, A.N.; Barta, J.R.; Smith, D.A. *Babesia odocoilei* as a cause of mortality in captive cervids in Canada. *Can. Vet. J.* **2018**, *59*, 52–58. [PubMed]
12. Luppi, M.M.; Malta, M.C.; Silva, T.M.; Silva, F.L.; Motta, R.O.; Miranda, I.; Ecco, R.; Santos, R.L. Visceral leishmaniasis in captive wild canids in Brazil. *Vet. Parasitol.* **2008**, *155*, 146–151. [CrossRef] [PubMed]
13. Fleischman, D.A.; Chomel, B.B.; Kasten, R.W.; André, M.R.; Gonçalves, L.R.; Machado, R.Z. *Bartonella clarridgeiae* and *Bartonella vinsonii* subsp. *berkhoffii* exposure in captive wild canids in Brazil. *Epidemiol. Infect.* **2015**, *143*, 573–577. [CrossRef] [PubMed]
14. Silveira, J.A.; Rabelo, E.M.; Ribeiro, M.F. Molecular detection of tick-borne pathogens of the family Anaplasmataceae in Brazilian brown brocket deer (*Mazama gouazoubira*, Fischer, 1814) and marsh deer (*Blastocerus dichotomus*, Illiger, 1815). *Transbound. Emerg. Dis.* **2012**, *59*, 353–360. [CrossRef] [PubMed]
15. Yamamoto, K.; Chomel, B.B.; Lowenstine, L.J.; Kikuchi, Y.; Phillips, L.G.; Barr, B.C.; Swift, P.K.; Jones, K.R.; Riley, S.P.D.; Kasten, R.W.; et al. *Bartonella henselae* antibody prevalence in free-ranging and captive wild felids from California. *J. Wildl. Dis.* **1998**, *34*, 56–63. [CrossRef]
16. Millán, J.; Di Cataldo, S.; Volokhov, D.V.; Becker, D.J. Worldwide occurrence of hemoplasmas in wildlife: Insights into the patterns of infection, transmission, pathology, and zoonotic potential. *Transbound. Emerg. Dis.* **2020**, *68*, 3236–3256. [CrossRef]
17. Prefeitura de Belo Horizonte; Fundação de Parques Municipais e Zoobotânica. Belo Horizonte. PBH. Available online: https://prefeitura.pbh.gov.br/fundacao-de-parques-e-zoobotanica (accessed on 15 November 2024).
18. Birkenheuer, A.J.; Levy, M.G.; Breitschwerdt, E.B. Development and evaluation of a seminested PCR for detection and differentiation of *Babesia gibsoni* (*Asian genotype*) and *B. canis* DNA in canine blood samples. *J. Clin. Microbiol.* **2003**, *41*, 4172–4177. [CrossRef]
19. Massung, R.F.; Slater, K.; Owens, J.H.; Nicholson, W.L.; Mather, T.N.; Solberg, V.B.; Olson, J.G. Nested PCR assay for detection of granulocytic ehrlichiae. *J. Clin. Microbiol.* **1998**, *36*, 1090–1095. [CrossRef]
20. Kawahara, M.; Tajima, T.; Torii, H.; Yabutani, M.; Ishii, J.; Harasawa, M.; Isogai, E.; Rikihisa, Y. *Ehrlichia chaffeensis* infection of sika deer, Japan. *Emerg. Infect. Dis.* **2009**, *15*, 1991–1993. [CrossRef]

21. De la Fuente, J.; Massung, R.F.; Wong, S.J.; Chu, F.K.; Lutz, H.; Meli, M.; von Loewenich, F.D.; Grzeszczuk, A.; Torina, A.; Caracappa, S.; et al. Sequence analysis of the msp4 gene of *Anaplasma phagocytophilum* strains. *J. Clin. Microbiol.* **2005**, *43*, 1309–1317. [CrossRef]
22. Bown, K.J.; Lambin, X.; Ogden, N.H.; Petrovec, M.; Shaw, S.E.; Woldehiwet, Z.; Birtles, R.J. High-resolution genetic fingerprinting of European strains of *Anaplasma phagocytophilum* by use of multilocus variable-number tandem-repeat analysis. *J. Clin. Microbiol.* **2007**, *45*, 1771–1776. [CrossRef] [PubMed]
23. De la Fuente, J.; Ruiz-Fons, F.; Naranjo, V.; Torina, A.; Rodríguez, O.; Gortázar, C. Evidence of *Anaplasma* infections in European roe deer (*Capreolus capreolus*) from southern Spain. *Res. Vet. Sci.* **2008**, *84*, 382–386. [CrossRef] [PubMed]
24. Ramos, C.A.N.; Ramos, R.A.N.; Araújo, F.R.; Júnior, D.S.G.; Souza, I.I.F.; Ono, T.M.; Pimentel, D.S.; Rosas, E.O.; Faustino, M.A.G.; Alves, L.C. Comparação de nested-PCR com o diagnóstico direto na detecção de Ehrlichia canis e Anaplasma platys em cães. *Rev. Bras. Parasitol. Vet.* **2009**, *18*, 58–62. [CrossRef] [PubMed]
25. Zahler, M.; Rinder, H.; Schein, E.; Gothe, R. Detection of a new pathogenic Babesia microti-like species in dogs. *Vet. Parasitol.* **2000**, *89*, 241–248. [CrossRef]
26. Silveira, J.A.; Rabelo, E.M.; Ribeiro, M.F. Detection of Theileria and Babesia in brown brocket deer (*Mazama gouazoubira*) and marsh deer (*Blastocerus dichotomus*) in the State of Minas Gerais, Brazil. *Vet. Parasitol.* **2011**, *177*, 61–66. [CrossRef]
27. Birkenheuer, A.J.; Marr, H.; Alleman, A.R.; Levy, M.G.; Breitschwerdt, E.B. Development and evaluation of a PCR assay for the detection of Cytauxzoon felis DNA in feline blood samples. *Vet. Parasitol.* **2006**, *137*, 144–149. [CrossRef]
28. Schönian, G.; Nasereddin, A.; Dinse, N.; Schweynoch, C.; Schallig, H.D.; Presber, W.; Jaffe, C.L. PCR diagnosis and characterization of *Leishmania* in local and imported clinical samples. *Diagn. Microbiol. Infect. Dis.* **2003**, *47*, 349–358. [CrossRef]
29. Jensen, W.A.; Fall, M.Z.; Rooney, J.; Kordick, D.L.; Breitschwerdt, E.B. Rapid identification and differentiation of *Bartonella* species using a single-step PCR assay. *J. Clin. Microbiol.* **2000**, *38*, 1717–1722. [CrossRef]
30. Criado-Fornelio, A.; Martinez-Marcos, A.; Buling-Saraña, A.; Barba-Carretero, J.C. Presence of *Mycoplasma haemofelis*, *Mycoplasma haemominutum* and piroplasmids in cats from southern Europe: A molecular study. *Vet. Microbiol.* **2003**, *93*, 307–317. [CrossRef]
31. Bastos, C.; Passos, L.; Facury-Filho, E.; Rabelo, E.; de la Fuente, J.; Ribeiro, M. Protection in the absence of exclusion between two Brazilian isolates of *Anaplasma marginale* in experimentally infected calves. *Vet. J.* **2010**, *186*, 374–378. [CrossRef]
32. Moreira, S.M. Estudo retrospectivo (1998–2001) da erliquiose canina em Belo Horizonte: Avaliação clínica e laboratorial de infecções experimentais. Master's Thesis, Universidade Federal de Minas Gerais, Belo Horizonte, Minas Gerais, Brazil, 2001. Available online: http://hdl.handle.net/1843/BUOS-8C8DX2 (accessed on 15 November 2024).
33. Costa, V.M.d.M.; Ribeiro, M.F.B.; Duarte, A.L.L.; Mangueira, J.M.; Pessoa, A.F.A.; Azevedo, S.S.; de Barros, A.T.M.; Riet-Correa, F.; Labruna, M.B. Seroprevalence and risk factors for cattle anaplasmosis, babesiosis, and trypanosomiasis in a Brazilian semiarid region. *Rev. Bras. Parasitol. Vet.* **2013**, *22*, 207–213. [CrossRef] [PubMed]
34. Castillo, A.P.; Miranda, J.V.O.; Fonseca, P.L.C.; Silva, S.d.O.; Lopes, R.E.N.; Spanhol, V.C.; Moreira, R.G.; Nicolino, R.R.; Queiroz, D.C.; Santos, L.C.G.d.A.e.; et al. Evidence of SARS-CoV-2 infection and co-infections in stray cats in Brazil. *Acta Trop.* **2024**, *249*, 107056. [CrossRef] [PubMed]
35. Hall, T.A. BioEdit: A user-friendly biological sequence alignment editor and analysis program for Windows 95/98/NT. *Nucleic Acids Symp. Ser.* **1999**, *41*, 95–98. Available online: https://www.academia.edu/2034992/BioEdit_a_user_friendly_biological_sequence_alignment_editor_and_analysis_program_for_Windows_95_98_NT (accessed on 15 November 2024).
36. Altschul, S.F.; Gish, W.; Miller, W.; Myers, E.W.; Lipman, D.J. Basic local alignment search tool. *J. Mol. Biol.* **1990**, *215*, 403–410. [CrossRef]
37. Katoh, K.; Standley, D.M. MAFFT multiple sequence alignment software version 7: Improvements in performance and usability. *Mol. Biol. Evol.* **2013**, *30*, 772–780. [CrossRef]
38. Nei, M.; Kumar, S. *Molecular Evolution and Phylogenetics*; Oxford University Press: New York, NY, USA, 2000.
39. Kimura, M. A simple method for estimating evolutionary rate of base substitutions through comparative studies of nucleotide sequences. *J. Mol. Evol.* **1980**, *16*, 111–120. [CrossRef]
40. Tamura, K.; Stecher, G.; Kumar, S. MEGA11: Molecular Evolutionary Genetics Analysis Version 11. *Mol. Biol. Evol.* **2021**, *38*, 3022–3027. [CrossRef]
41. Jukes, T.H.; Cantor, C.R. Evolution of Protein Molecules. In *Mammalian Protein Metabolism*; Munro, H., Ed.; Academic Press: New York, NY, USA, 1969; Volume III, Charpter 24; pp. 21–132.
42. Felsenstein, J. Confidence limits on phylogenies: An approach using the bootstrap. *Evol. Int. J. Org. Evol.* **1985**, *39*, 783–791. [CrossRef]
43. Robinette, C.; Saffran, L.; Ruple, A.; Deem, S. Zoos and public health: A partnership on the One Health frontier. *One Health* **2016**, *3*, 1–4. [CrossRef]
44. Silveira, J.A.; Valente, P.C.; Paes, P.R.; Vasconcelos, A.V.; Silvestre, B.T.; Ribeiro, M.F. The first clinical and laboratory evidence of co-infection by *Anaplasma phagocytophilum* and *Ehrlichia canis* in a Brazilian dog. *Ticks Tick-Borne Dis.* **2015**, *6*, 242–245. [CrossRef]
45. Lasta, C.S.; dos Santos, A.P.; Messick, J.B.; Oliveira, S.T.; Biondo, A.W.; Vieira, R.F.d.C.; Dalmolin, M.L.; González, F.H.D. Molecular detection of *Ehrlichia canis* and *Anaplasma platys* in dogs in Southern Brazil. *Rev. Bras. Parasitol. Vet.* **2013**, *22*, 360–366. [CrossRef] [PubMed]
46. De Arruda, R.M.F.; Cardoso, D.T.; Teixeira-Neto, R.G.; Barbosa, D.S.; Ferraz, R.K.; Morais, M.H.F.; Belo, V.S.; da Silva, E.S. Space-time analysis of the incidence of human visceral leishmaniasis (VL) and prevalence of canine VL in a municipality of southeastern Brazil: Identification of priority areas for surveillance and control. *Acta Trop.* **2019**, *197*, 105052. [CrossRef] [PubMed]

47. Lopes, J.V.; Michalsky, M.; Pereira, N.C.L.; de Paula, A.J.V.; Souza, A.G.M.; Pinheiro, L.C.; Lima, A.C.V.M.d.R.; de Avelar, D.M.; França-Silva, J.C.; Lanzetta, V.A.S.; et al. Canine visceral leishmaniasis in area with recent Leishmania transmission: Prevalence, diagnosis, and molecular identification of the infecting species. *Rev. Soc. Bras. Med. Trop.* **2020**, *53*, e20200141. [CrossRef] [PubMed]
48. Xavier, S.; Chiarelli, I.; Lima, W.; Gonçalves, R.; Tafuri, W. Canine visceral leishmaniasis: A remarkable histopathological picture of one asymptomatic animal reported from Belo Horizonte, Minas Gerais, Brazil. *Arq. Bras. Med. Vet. Zootec.* **2006**, *58*, 944–1000. [CrossRef]
49. De Fraga, A.P.; da Silveira, V.P.; Salla, P.d.F.; Goulart, F.G.d.O.; Streck, A.F.; Pereira, V.R.Z.B.; de Mello, L.S.; Fonseca, A.S.K.; Ikuta, N.; Lunge, V.R. Canine Leishmaniasis in Southern Brazil: Diagnosis and Clinical Features in Domestic Dogs. *Zoonotic Dis.* **2024**, *4*, 114–122. [CrossRef]
50. André, M.R.; Denardi, N.C.B.; de Sousa, K.C.M.; Gonçalves, L.R.; Henrique, P.C.; Ontivero, C.R.G.R.; Gonzalez, I.H.L.; Nery, C.V.C.; Chagas, C.R.F.; Monticelli, C.; et al. Arthropod-borne pathogens circulating in free-roaming domestic cats in a zoo environment in Brazil. *Ticks Tick-Borne Dis.* **2014**, *5*, 545–551. [CrossRef]
51. Calchi, A.C.; Yogui, D.R.; Alves, M.H.; Desbiez, A.L.J.; Kluyber, D.; Vultão, J.G.; Arantes, P.V.C.; de Santi, M.; Werther, K.; Teixeira, M.M.G.; et al. Molecular detection of piroplasmids in mammals from the Superorder Xenarthra in Brazil. *Parasitol. Res.* **2023**, *122*, 3169–3180. [CrossRef]
52. Fava, N.M.N.; Alves, T.S.; Lopes, M.G.; Labruna, M.B.; Santos, A.Q.; Cury, M.C. Occurrence and Molecular Identification of *Hemoparasites* in Wild Mammals Kept in Rehabilitation Centers in Brazil. *Acta Parasitol.* **2022**, *67*, 476–486. [CrossRef]
53. Furtado, M.M.; Taniwaki, S.A.; Metzger, B.; Paduan, K.d.S.; O'dwyer, H.L.; Jácomo, A.T.d.A.; Porfírio, G.E.; Silveira, L.; Sollmann, R.; Tôrres, N.M.; et al. Is the free-ranging jaguar (*Panthera onca*) a reservoir for *Cytauxzoon felis* in Brazil? *Ticks Tick-Borne Dis.* **2017**, *8*, 470–476. [CrossRef]
54. McNew, S.M.; Barrow, L.N.; Williamson, J.L.; Galen, S.C.; Skeen, H.R.; DuBay, S.G.; Gaffney, A.M.; Johnson, A.B.; Bautista, E.; Ordoñez, P.; et al. Contrasting drivers of diversity in hosts and parasites across the tropical Andes. *Proc. Natl. Acad. Sci. USA* **2021**, *118*, e2010714118. [CrossRef]
55. Otiende, M.Y.; Kivata, M.W.; Makumi, J.N.; Mutinda, M.N.; Okun, D.; Kariuki, L.; Obanda, V.; Gakuya, F.; Mijele, D.; Soriguer, R.C.; et al. Epidemiology of *Theileria bicornis* among black and white rhinoceros metapopulation in Kenya. *BMC Vet. Res.* **2015**, *11*, 4. [CrossRef] [PubMed]
56. Silva, V.L.d.B.; de Almeida, S.L.H.; Maia, M.O.; Santos, T.; Pavelegini, L.A.D.; Zaffalon, G.B.; Marcili, A.; Morgado, T.O.; Dutra, V.; Nakazato, L.; et al. Post mortem protozoan hemoparasites detection in wild mammals from Mato Grosso state, Midwestern Brazil. *Rev. Bras. Parasitol. Vet.* **2021**, *30*, e013021. [CrossRef] [PubMed]
57. Guizelini, C.C.; Nascimento, C.A.; Echeverria, J.T.; Soares, R.L.; Pimenta, M.M.; de Deco-Souza, T.; Esteves, F.C.; Gomes, D.C. Fatal infection caused by *Cytauxzoon felis* in a captive-reared jaguar (*Panthera onca*). *Int. J. Parasitol. Parasites Wildl.* **2021**, *16*, 187–190. [CrossRef] [PubMed]
58. Fenelon, A.C.G.; da Hora, A.S.; da Silva, K.L.; de Oliveira, G.H.B.; Gonçalves, M.S.e.S.; Pastor, F.M.; Barbosa, F.C.; Siqueira, M.T.S.; Rosalinski-Moraes, F. Co-infection of *Cytauxzoon felis*, *Mycoplasma haemofelis*, and the feline immunodeficiency virus in a domestic cat in Uberlândia, Minas Gerais, Brazil. *Braz. J. Vet. Res. Anim. Sci.* **2023**, *60*, e210131. [CrossRef]
59. Ikeda, P.; Menezes, T.R.; Torres, J.M.; de Oliveira, C.E.; Lourenço, E.C.; Herrera, H.M.; Machado, R.Z.; André, M.R. First molecular detection of piroplasmids in non-hematophagous bats from Brazil, with evidence of putative novel species. *Parasitol. Res.* **2021**, *120*, 301–310. [CrossRef]
60. Yam, J.; Gestier, S.; Bryant, B.; Campbell-Ward, M.; Bogema, D.; Jenkins, C. The identification of *Theileria bicornis* in captive rhinoceros in Australia. *Int. J. Parasitol. Parasites Wildl.* **2017**, *7*, 85–89. [CrossRef]
61. Otiende, M.Y.; Kivata, M.W.; Jowers, M.J.; Makumi, J.N.; Runo, S.; Obanda, V.; Gakuya, F.; Mutinda, M.; Kariuki, L.; Alasaad, S. Three Novel Haplotypes of *Theileria bicornis* in Black and White Rhinoceros in Kenya. *Transbound. Emerg. Dis.* **2016**, *63*, e144–e150. [CrossRef]
62. Sivakumar, T.; Fujita, S.; Tuvshintulga, B.; Kothalawala, H.; Silva, S.S.P.; Yokoyama, N. Discovery of a new *Theileria* sp. closely related to *Theileria annulata* in cattle from Sri Lanka. *Sci. Rep.* **2019**, *9*, 16132. [CrossRef]
63. Silveira, J.A.G.; D'Elia, M.L.; Avelar, I.d.O.; de Almeida, L.R.; dos Santos, H.A.; Soares, D.F.d.M.; Ribeiro, M.F.B.; Lima, W.d.S.; Ecco, R. *Rangelia vitalii* in a free-ranging maned wolf (*Chrysocyon brachyurus*) and co-infections. *Int. J. Parasitol. Parasites Wildl.* **2016**, *5*, 280–285. [CrossRef]
64. Mol, J.P.; Soave, S.A.; Turchetti, A.P.; Pinheiro, G.R.; Pessanha, A.T.; Malta, M.C.; Tinoco, H.P.; Figueiredo, L.A.; Gontijo, N.F.; Paixão, T.A.; et al. Transmissibility of *Leishmania infantum* from maned wolves (*Chrysocyon brachyurus*) and bush dogs (*Speothos venaticus*) to *Lutzomyia longipalpis*. *Vet. Parasitol.* **2015**, *212*, 86–91. [CrossRef]
65. Schäfer, I.; Kohn, B.; Volkmann, M.; Müller, E. Retrospective evaluation of vector-borne pathogens in cats living in Germany (2012–2020). *Parasites Vectors* **2021**, *14*, 123. [CrossRef] [PubMed]
66. Pereira, A.; Parreira, R.; Nunes, M.; Casadinho, A.; Vieira, M.L.; Campino, L.; Maia, C. Molecular detection of tick-borne bacteria and protozoa in cervids and wild boars from Portugal. *Parasites Vectors* **2016**, *9*, 251. [CrossRef] [PubMed]
67. Kolo, A.O.; Collins, N.E.; Brayton, K.A.; Chaisi, M.; Blumberg, L.; Frean, J.; Gall, C.A.; Wentzel, J.M.; Wills-Berriman, S.; De Boni, L.; et al. *Anaplasma phagocytophilum* and Other *Anaplasma* spp. in Various Hosts in the Mnisi Community, Mpumalanga Province, South Africa. *Microorganisms* **2020**, *8*, 1812. [CrossRef] [PubMed]

68. Salazar, A.; Ochoa-Corona, F.M.; Talley, J.L.; Noden, B.H. Recombinase polymerase amplification (RPA) with lateral flow detection for three *Anaplasma* species of importance to livestock health. *Sci. Rep.* **2021**, *11*, 15962. [CrossRef]
69. Gomes, P.D.; Hirano, L.Q.L.; de Paula, R.C. Epidemiological survey of infectious agents in free-ranging maned wolves (*Chrysocyon brachyurus*) in Northeastern Brazil. *Braz. J. Microbiol.* **2024**, *55*, 933–941. [CrossRef]
70. Nasr, A.; Ghafar, M.; El hariri, M. Detection of *Anaplasma platys* and *Ehrlichia canis* in *Rhipicephalus sanguineus* ticks attached to dogs from Egypt; a public health concern. *Vet. Med. J.* **2020**, *66*, 1–9. [CrossRef]
71. Luo, Y.Y.; Yu, D.; Zhang, H.Z.; Liu, Z.X.; Hong, R.D.; Hong, M.; Ai, Z.Q.; Zhu, J.J.; Yin, J.X. Molecular detection of *Bartonella* species in wild small mammals in western Yunnan Province, China. *Front. Vet. Sci.* **2023**, *10*, 1301316. [CrossRef]
72. Mongruel, A.C.B.; Medici, E.P.; Canena, A.d.C.; Dias, C.M.; Machado, R.Z.; André, M.R. Molecular evidence of *Bartonella* spp. in wild lowland tapirs (*Tapirus terrestris*), the largest land mammals in Brazil. *Comp. Immunol. Microbiol. Infect. Dis.* **2023**, *101*, 102042. [CrossRef]
73. Rao, H.; Li, S.; Lu, L.; Wang, R.; Song, X.; Sun, K.; Shi, Y.; Li, D.; Yu, J. Genetic diversity of Bartonella species in small mammals in the Qaidam Basin, western China. *Sci. Rep.* **2021**, *11*, 1735. [CrossRef]
74. Greco, G.; Zarea, A.A.K.; Sgroi, G.; Tempesta, M.; D'alessio, N.; Lanave, G.; Bezerra-Santos, M.A.; Iatta, R.; Veneziano, V.; Otranto, D.; et al. Zoonotic Bartonella species in Eurasian wolves and other free-ranging wild mammals from Italy. *Zoonoses Public Health* **2021**, *68*, 316–326. [CrossRef]
75. Pacheco, T.d.A.; Amaral, R.B.D.; Ikeda, P.; Maia, M.O.; Lee, D.A.B.; Semedo, T.B.F.; de Mendonça, R.F.B.; Pedroni, F.; Horta, M.C.; Rossi, R.V.; et al. Molecular detection and characterization of *Bartonella* spp. in small mammals in the Amazonia and Cerrado biomes, midwestern Brazil. *Acta Trop.* **2024**, *251*, 107129. [CrossRef] [PubMed]
76. Cubilla, M.P.; Santos, L.C.; de Moraes, W.; Cubas, Z.S.; Leutenegger, C.M.; Estrada, M.; Vieira, R.F.; Soares, M.J.; Lindsay, L.L.; Sykes, J.E.; et al. Occurrence of hemotropic mycoplasmas in non-human primates (*Alouatta caraya*, *Sapajus nigritus* and *Callithrix jacchus*) of southern Brazil. *Comp. Immunol. Microbiol. Infect. Dis.* **2017**, *52*, 6–13. [CrossRef] [PubMed]
77. Mongruel, A.C.B.; Somma, A.T.; Pinto, A.C.A.; Campos, C.d.F.; Calado, M.I.Z.; Montiani-Ferreira, F.; Vieira, T.S.W.J.; Vieira, R.F.d.C. First Report of 'Candidatus Mycoplasma haematomacacae' in Laboratory-Kept Rhesus Monkeys (*Macaca mulatta*) Maintained in Rio de Janeiro, Brazil. *Vet. Sci.* **2022**, *9*, 443. [CrossRef] [PubMed]
78. De Melo, C.M.F.; Daneze, E.R.; Mendes, N.S.; Ramos, I.A.d.S.; Morales-Donoso, J.A.; Fernandes, S.J.; Machado, R.Z.; André, M.R.; Sobreira, M.F.d.R. Genetic diversity and hematological and biochemical alterations in *Alouatta primates* naturally infected with hemoplasmas in Brazil. *Comp. Immunol. Microbiol. Infect. Dis.* **2019**, *63*, 104–111. [CrossRef]
79. Carneiro, F.T.; Scalon, M.C.; Amorim, G.; Silva, W.A.; Honorato, S.M.; Pereira, F.M.; Silva, L.H.; Aquino, L.C.; Paludo, G.R. Hemoplasma infection in wild captive carnivores. *Pesqui. Vet. Bras.* **2020**, *40*, 293–299. [CrossRef]
80. André, M.R.; Duarte, J.M.B.; Gonçalves, L.R.; Sacchi, A.B.V.; Jusi, M.M.G.; Machado, R.Z. New records and genetic diversity of *Mycoplasma ovis* in free-ranging deer in Brazil. *Epidemiol. Infect.* **2020**, *148*, e6. [CrossRef]
81. Mongruel, A.C.B.; Medici, E.P.; Canena, A.D.C.; Calchi, A.C.; Machado, R.Z.; André, M.R. Expanding the Universe of Hemoplasmas: Multi-Locus Sequencing Reveals Putative Novel Hemoplasmas in Lowland Tapirs (*Tapirus terrestris*), the Largest Land Mammals in Brazil. *Microorganisms* **2022**, *10*, 614. [CrossRef]
82. Andrews, K.R.; Besser, T.E.; Stalder, T.; Top, E.M.; Baker, K.N.; Fagnan, M.W.; New, D.D.; Schneider, G.M.; Gal, A.; Andrews-Dickert, R.; et al. Comparative genomic analysis identifies potential adaptive variation and virulence factors in *Mycoplasma ovipneumoniae*. *bioRxiv* **2024**. [CrossRef]
83. De Mello, V.V.C.; Calchi, A.C.; de Oliveira, L.B.; Coelho, T.F.S.B.; Lee, D.A.B.; Franco, E.O.; Machado, R.Z.; André, M.R. Molecular Survey of Piroplasmids and Hemosporidians in Vampire Bats, with Evidence of Distinct Piroplasmida Lineages Parasitizing *Desmodus rotundus* from the Brazilian Amazon. *Parasitologia* **2023**, *3*, 248–259. [CrossRef]
84. Johnson, J.S.; Spakowicz, D.J.; Hong, B.-Y.; Petersen, L.M.; Demkowicz, P.; Chen, L.; Leopold, S.R.; Hanson, B.M.; Agresta, H.O.; Gerstein, M.; et al. Evaluation of 16S rRNA gene sequencing for species and strain-level microbiome analysis. *Nat. Commun.* **2019**, *10*, 5029. [CrossRef]
85. Kuhn, J.; Marti, I.; Ryser-Degiorgis, M.-P.; Wernike, K.; Jones, S.; Tyson, G.; Delalay, G.; Scherrer, P.; Borel, S.; Hosie, M.J.; et al. Investigations on the Potential Role of Free-Ranging Wildlife as a Reservoir of SARS-CoV-2 in Switzerland. *Viruses* **2024**, *16*, 1407. [CrossRef] [PubMed]
86. Young, A.D.; Gillung, J.P. Phylogenomics-principles, opportunities and pitfalls of big-data phylogenetics. *Syst. Entomol.* **2020**, *45*, 225–247. [CrossRef]
87. Firouzeh, N.; Borj, H.F.; Ziaali, N.; Kareshk, A.T.; Ahmadinejad, M.; Shafiei, R. Genetic Diversity of *Toxoplasma gondii* by Serological and Molecular Analyzes in Different Sheep and Goat Tissues in Northeastern Iran. *Iran. J. Parasitol.* **2023**, *18*, 217–228. [CrossRef] [PubMed]
88. Shafiei, R.; Firouzeh, N.; Rahimi, M.T. Serological and molecular survey of *Toxoplasma gondii* in aborted livestock fetuses from Northeast Iran. *BMC Res. Notes* **2024**, *17*, 290. [CrossRef]

Journal of
Zoological and Botanical Gardens

MDPI

Article

A New Soil-Based Solution for Reducing the Persistence of Parasites in Zoos

José Ángel Hernández [1,2,*,†], Cándido Viña [1], João Lozano [3,4], Izaro Zubiría [1], Miguel Baena [1], Inês Abreu-Ramos [1], Rami Salmo [1], Esther Valderrábano [1,5], Gustavo Pérez-Anzúrez [6,7], Adolfo Paz-Silva [1], Rodrigo Muíño [2,†], Cristiana Cazapal-Monteiro [1], María Sol Arias [1] and Mercedes Camiña [8,†]

1 Control of Parasites Group (COPAR, GI-2120), Department of Animal Pathology, Faculty of Veterinary, University of Santiago de Compostela, 27002 Lugo, Spain; candidovina@hotmail.es (C.V.); izaro89@gmail.com (I.Z.); miguel.baena.laseca@rai.usc.es (M.B.); inesabreu.ramos@usc.es (I.A.-R.); rami.salmo22@gmail.com (R.S.); esthervalcano@gmail.com (E.V.); adolfo.paz@usc.es (A.P.-S.); cristiana.cazapal@usc.es (C.C.-M.); mariasol.arias@usc.es (M.S.A.)

2 Propaedeutic Clinic, Department of Animal Pathology, Faculty of Veterinary, University of Santiago de Compostela, 27002 Lugo, Spain; rodrigo.muino.otero@usc.es

3 CIISA—Centre for Interdisciplinary Research in Animal Health, Faculty of Veterinary Medicine, University of Lisbon, Avenida da Universidade Técnica, 1300-477 Lisbon, Portugal; jlozano@fmv.ulisboa.pt

4 Associate Laboratory for Animal and Veterinary Sciences (AL4AnimalS), 1300-477 Lisbon, Portugal

5 Marcelle Natureza Zoological Park, 27154 Outeiro de Rei, Spain

6 Laboratory of Helminthology, National Centre for Disciplinary Research in Animal Health and Innocuity (CENID-SAI), National Institute for Research in Forestry, Agriculture and Livestock-Ministry of Agriculture and Rural Development, (INIFAP-SADER), Jiutepec 62550, Mexico; tavopzaz@gmail.com

7 Production Sciences and Animal Health, Faculty of Veterinary Medicine and Zootechnics, National Autonomous University of Mexico, Coyoacán, Ciudad de Mexico 04510, Mexico

8 Department of Physiology, Faculty of Veterinary, University of Santiago de Compostela, 27002 Lugo, Spain; merchi.camina@usc.es

* Correspondence: joseangel.malagon@usc.es

† These authors contributed equally to this work.

Abstract: Controlling parasites in zoo animals is a significant challenge, making practical infection prevention methods essential. This study evaluated a novel solution using parasitophagous fungi-treated forage seeds to reduce soil parasite persistence. We conducted two experiments to assess the soil distribution of the fungi *Mucor circinelloides* (ovicidal) and *Duddingtonia flagrans* (larvicidal). Forage seeds were immersed in a submerged culture containing 10^6 spores/mL and subsequently sown in plastic trays (trial 1) and demarcated ground plots (40 × 30 cm) framed with wood (trial 2). Fifteen days later, *Ascaris suum* and cyathostomin eggs were placed above the germinated vegetation. After a 25–30-day period, the viability of roundworm eggs reduced by 62%, and half of them remained undeveloped; a 64% reduction in the counts of L3 cyathostomins was recorded. In trial 2, viability decreased by 55% in roundworm eggs, with an infectivity rate of 30%, while L3 counts lessened by 57%. It has been concluded that the risk of infection by ascarids and cyathostomins can be reduced by at least half by sowing the soil with forage seeds pre-treated with fungi, providing a practical solution for parasite control in zoos.

Keywords: helminths; zoological park; fungal-treated seeds; *Mucor circinelloides*; *Duddingtonia flagrans*

1. Introduction

The initial purpose of zoos was to exhibit animals for display as a kind of museum, preserving exotic specimens obtained on expeditions. Over time, there have been notable changes, and zoos now focus on conservation as wild habitats continue diminishing, and many animals are at greater risk of extinction [1]. In addition, activities are often organized to promote education among visitors and to participate in research projects, always following the highest standards of animal care and welfare [2].

Modern zoos strive to conserve a diverse array of animal species by maintaining them in habitats that closely resemble their natural ecosystems. For example, all enclosures for bears should contain a dry resting area, pool, and den [3]. Zones large enough to allow wolves to chase each other are suggested in zoological gardens, whereas small felids like lynxes or bobcats in captivity need perching platforms at or near the top of their enclosure, a place from which they can "hide" and peer out. They also require logs upon which they can "sharpen" their claws [4]. In constructing a shorebird habitat, a mixture of ground cover that includes sand, rocks, clay, grass mats, and saltwater is suggested. Driftwood and rocks—which they would find in their natural environment—serve as perfect "furniture" for them to perch on. There is also running water in their habitats, and the birds seem to enjoy wading and standing at the water's edge. Regarding herbivores, significant efforts have been made in many zoos to prepare grassy plots in order to offer them the possibility to nourish, socialize, and interact with the environment [5].

Some of the measures introduced in recent years in zoos have improved animal welfare, but other problems remain unresolved and even seem to have worsened, such as the control of certain parasitic infections [6]. Providing herbivores with grassy enclosures replicates their natural habitats but also facilitates the lifecycle completion of strongylid nematodes, potentially leading to significant re-infections [7,8]. On the other hand, the presence of soil in the plots where carnivores are kept makes it easier for geohelminths such as ascarids or trichurids to finish their cycle until the respective infective stages are reached. Therefore, the situation becomes similar to that experienced by domestic animals, and deworming as the only measure reveals insufficient to obtain good results so that animals can reinfect often and quite rapidly [9], which reinforces the need to have complementary procedures to support the deworming.

Given recent efforts to restore natural environments, there is potential to decrease reliance on anthelmintics by focusing on soil-dwelling microorganisms that can disrupt or inhibit the development of parasitic stages pathogenic to animals [10,11]. In coincidence with this line, some experience acquired in the last decade underlines the possibility of using parasitophagous fungi for the control of parasitic forms affecting captive animals in zoos, focused mainly against parasites developing a direct life cycle as roundworms or strongylid nematodes [12]. Despite this information not being very abundant, different formulations involving the administration of saprophytic filamentous fungi such as *Mucor circinelloides* Van Thiegem (1875) and *Trichoderma atrobrunneum* F.B. Rocha, Chaverri & Jaklitsch (2015) (ovicidal), and *Duddingtonia flagrans* (Dudd.) R.C. Cooke (1969) (larvicidal) have been tried.

Given the satisfactory results obtained using various soil saprophytic filamentous fungi, a novel approach was developed to significantly reduce the prevalence of parasites in zoos. This strategy consists of soaking forage seeds in a submerged medium containing a mixture of chlamydospores of *Mucor circinelloides* and *Duddingtonia flagrans*, which are thick-walled structures that help fungi endure harsh conditions prior to sowing. The main objective is to restore natural soil conditions for wildlife by re-establishing the balance between hosts, pathogens, and antagonists. This method represents an innovative advance in the possibilities of biological parasite control, providing a practical and sustainable solution to improve the health and welfare of animals in zoological environments.

2. Materials and Methods

2.1. Parasitophagous Fungi

In the present investigation, a blend of two filamentous soil fungi (*M. circinelloides* CECT 20824 and *D. flagrans* CECT 20823) were co-cultured in the COPFr submerged medium to concentrations around 10^6 chlamydospores each per mL (Figure 1) [13].

These fungi were isolated by the COPAR Research Group (GI-2120; University of Santiago de Compostela, Spain) and deposited in the Spanish Type Culture Collection (CECT, Valencia, Spain) [13].

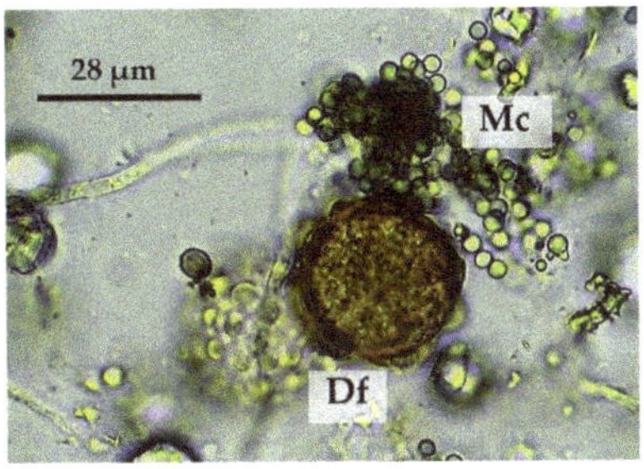

Figure 1. Chlamydospores of *M. circinelloides* (Mc) and *D. flagrans* (Df) were jointly grown in the COPFr submerged medium.

2.2. Preparation of Forage Seeds for Sowing

A commercial formulation containing forage seeds was utilized in the present research, composed of 40% *Lolium perenne* solen, 10% *Lolium multiflorum locobelo*, 40% *Lolium* x *boucheanum* Kunth, 5% *Trifolium repens huia*, and 5% *Trifolium pratense rozeta*.

In particular, pre-treatment of the seeds comprised placing them in trays at a 1:2 (mass/volume) ratio in a liquid culture medium containing the parasiticidal fungi at a proportion of 200 mL of medium per 100 g of seeds (Figure 2). A period of 2–5 h was allowed for the seeds to soak properly, and then they were stored in plastic bags until use, which was always performed within 7 days.

Figure 2. Forage seeds were soaked in a submerged culture containing chlamydospores of *M. circinelloides* (Mc) and *D. flagrans* (Df) and then kept in plastic bags until use in trials 1 (trays) and 2 (ground).

In trial 1, the preparation of the trays consisted of placing 200 g of compost autoclaved in each and then planting 50 g seeds. Accordingly, a total of 32 trays were prepared and maintained in the lab at RT (17–22 °C). Twelve trays were assigned untreated seeds (controls). Six of these trays received 12 g of feces from piglets, shedding 3450 eggs of *Ascaris suum* per gram of feces (EPG). The other six trays received 12 g of feces from horses, shedding 685 cyathostomin EPG. Twenty trays were sown with pre-treated seeds (Figure 3) and then sorted into two groups of 10 each. The group of G-TAs received 12 g of feces from piglets, shedding 3450 *A. suum* EPG, and the G-TCy group received 12 g of feces from horses, shedding 685 cyathostomin EPG. In each group, half of the trays were analyzed 15 days later, and the remainder was analyzed at the end of the assay (day 30).

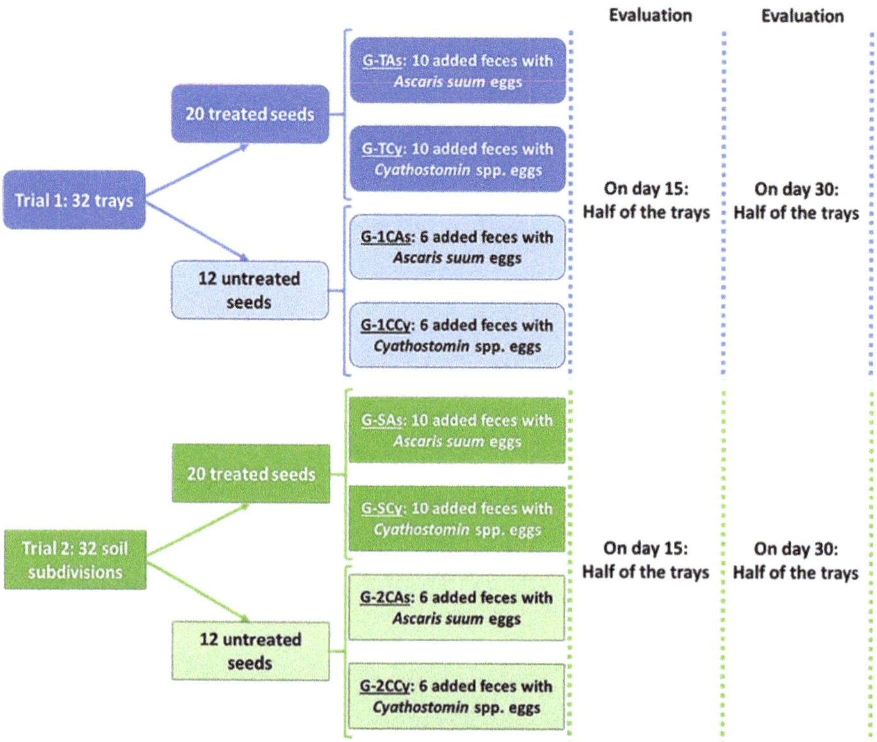

Figure 3. Study design conducted with forage seeds soaked in submerged culture containing chlamydospores of *M. circinelloides* (Mc) and *D. flagrans* (Df).

In trial 2, a total of 32 wooden frames (15 × 40 × 30 cm) were placed on the ground to delimit the test areas (Figure 2). The controls consisted of 12 sections sown with untreated seeds: six received 50 g of feces from piglets, shedding 3450 EPG *Ascaris suum*, and the other received six 50 g of feces from horses, shedding 685 cyathostomin EPG.

Twenty soil areas were sown with 100 g seeds pre-treated with fungi and then divided into two lots; G-SAs composed of 10 areas received 50 g of feces from piglets, shedding 3450 EPG *A. suum*, and another 10 received 50 g of feces from horses, shedding 685 cyathostomin EPG. Fifteen days later, half of the areas in each group were taken and analyzed, and the remaining were analyzed at the end of the assay (day 30).

2.3. Evaluation of the Antiparasitic Effect

In order to test the influence of the parasiticidal strategy, variations in the numbers of viable eggs of *A. suum* and of cyathostomin third-stage larvae (L3) were analyzed. This evaluation consisted of the observation of a minimum of 200 parasites (eggs or larvae, respectively) in each of the prepared trays and soil areas. Consequently, more than 10,000 observations of parasitic forms were performed throughout the study period using coprological probes.

2.3.1. Roundworms

With the aim of analyzing the antagonistic effect against roundworm eggs, a modified sedimentation technique [14] was used in trial 1, consisting of directly taking the content of each tray and emulsifying it into 1 L water. After completely mixing, the contents were sequentially passed through sieves with different pore diameters (4, 0.52, 0.15, and 0.050 mm) to reduce the presence of coarse particles. The resulting liquid was deposited in sedimentation cups for 24 h. Finally, the sediment obtained was analyzed under a microscope by placing 50 μL aliquots between glass slides and coverslips [15]. A similar procedure was applied to trial 2 by taking half of each soil sample delimited by the wooden frames (ca. 30 × 20 cm and 5 cm deep) and sinking it into 2 L water until entirely submerged. The blend was filtered along different sieves, as mentioned above, and the filtrate was placed into sedimentation cups for 24 h. After discarding the supernatant until 100 mL was obtained, the final volume was observed under a microscope, as mentioned before. Based on previous investigations, the eggs were sorted into non-viable and viable by considering the presence of eggshell damage or disruption [16] (Figure 4).

Viable eggs shed in feces	Infective stages developing in ground	Soil parasites exposed to filamentous fungi
A1 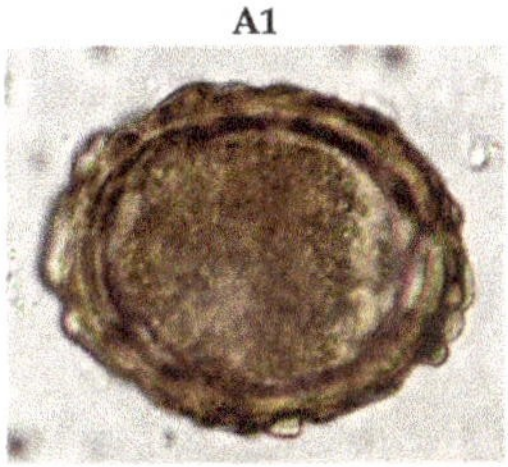	A2 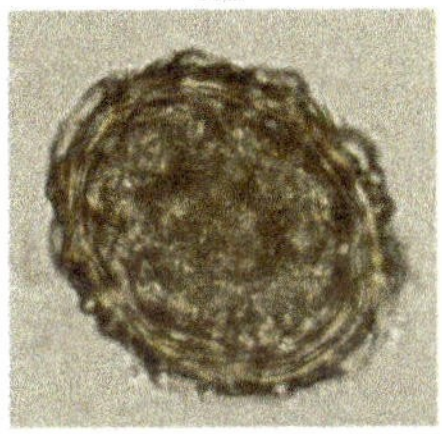	A3 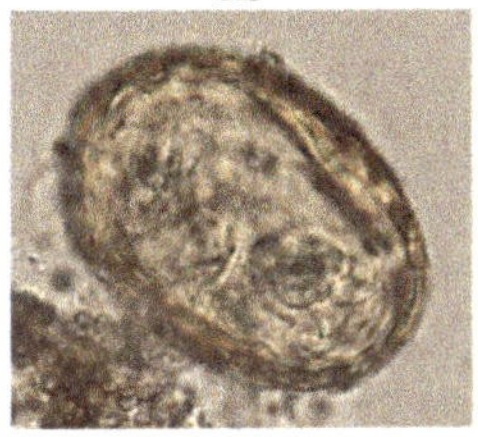
B1 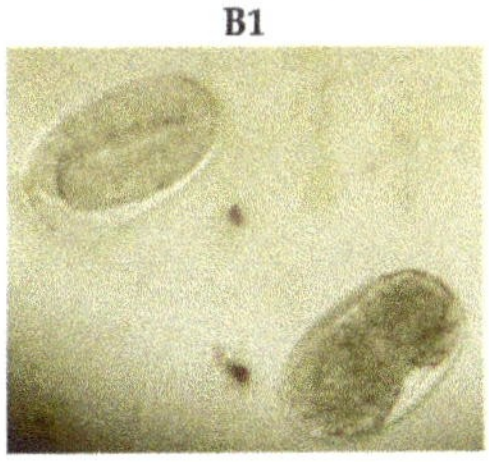	B2 	B3 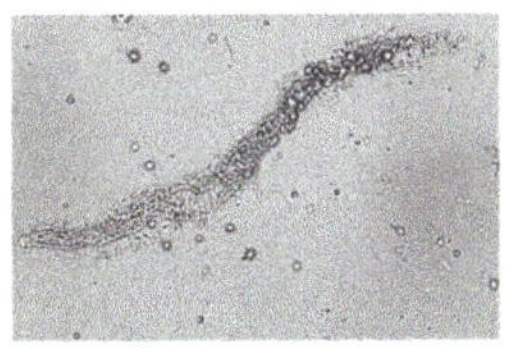

Figure 4. Eggs of certain helminths like *Ascaris sum* (**A1**) or strongylid nematodes (**B1**) are shed in the feces of infected animals and, once in the ground, evolve to reach the infective stage; eggs containing an L2 inside (**A2**) or third-stage larvae (**B2**). The antagonistic activity of the fungi causes the non-viability of eggs (**A3**) and L3 larvae (**B3**).

The percentage of reduction (PR) in eggs of *A. suum* was calculated according to the counts of viable eggs per gram of feces (EPG) on days 15 and 30, as follows:

$$PR_{15}\ (\%) = [1 - (EPG_{day15}/EPG_{day0})] \times 100$$

$$PR_{30}\ (\%) = [1 - (EPG_{day30}/EPG_{day0})] \times 100$$

Nevertheless, by considering that a percentage of eggs of *A. suum* becomes naturally non-viable [11], the possible effect attributed to the antiparasitic fungi was adjusted by taking into account the difference between PR in the exposed and in the controls; then, the Adjusted percentage of reduction (%) was calculated as

$$APR\ (\%) = PR_{treated} - PR_{controls}$$

2.3.2. Cyathostomins

In trial 1, the action on cyathostomin third-stage larvae (L3s) was evaluated by placing half of the content of each tray into 1 L of water and mixing for 1 h. Then, the contents were successively filtered through sieves with different pore diameters (4, 0.52, 0.15, and 0.050 mm), and the final solution was analyzed by means of the Baermann test for the collection of L3s: the infective stages [17] (Figure 4). The same protocol was applied to trial 2 on half of the content of each ground area. The identification of L3 cyathostomins was conducted using morphological keys [15]. Finally, the reduction in cyathostomin L3 larvae (PR_{L3}) was estimated as follows:

$$PR_{L3}\ (\%) = [1 - (LPG_{treated}/PG_{controls})] \times 100$$

2.4. Statistical Analysis

The Kolmogorov–Smirnov probe showed that the data collected for the fecal egg counts of *A. suum* were normally distributed ($Z = 1.155$, $p = 0.138$), and the Levene test revealed that the variances were homogeneous (Statistic = 4.792, $p = 0.053$). The analysis of the numbers of cyathostomin third-stage larvae (L3s) revealed that these data were also normal ($Z = 0.866$, $p = 0.441$), and the variances were homogeneous (Statistic = 4.792, $p = 0.053$). Accordingly, an ANOVA with repeated measures was conducted at a significance level of $p < 0.05$. All the probes were employed using the statistical software SPSS, version 22 (IBM SPSS, Inc., Chicago, IL, USA).

3. Results

3.1. Effect on Eggs of A. suum

As given in Figure 5, the number of viable *A. suum* eggs maintained in trays with herbage (G-TAs) for 30 days showed slight variations in the controls (ca. 6–10%). The number of viable eggs placed in those trays sown with seeds previously soaked in a liquid medium containing spores belonging to the two fungal species (G-1CAs) decreased by 19.3% during the first 15 days with respect to the controls (not exposed to the antiparasitic fungi) ($F = 42.148$, $p = 0.023$). At the end of the study (day 30), the number of viable eggs was reduced by 68%, and these differences were also significant ($F = 37.130$, $p = 0.026$).

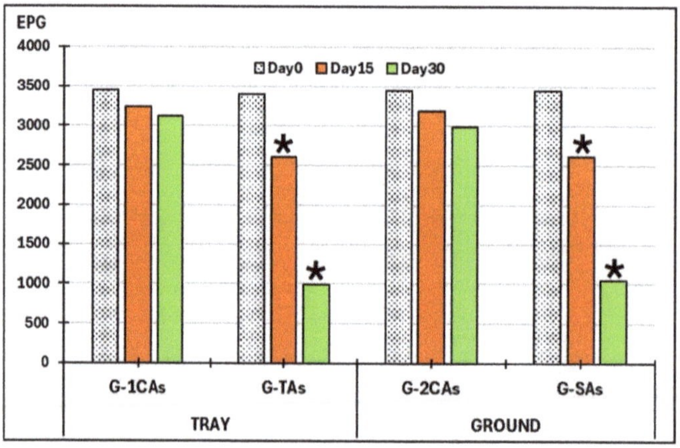

Figure 5. Variations in the viability of eggs of *Ascaris suum*. G-1CAs: control group trays (seeds without fungi); G-TAs: trays sown with seeds soaked in submerged culture containing *M. circinelloides* and *D. flagrans*; G-2CAs: control soil parcels (without fungi). G-SAs: soil parcels sown with seeds pre-soaked in submerged culture containing *M. circinelloides* and *D. flagrans*. EPG: eggs of *A. suum* per gram of feces; (*): significant differences.

Regarding data collected in control small ground areas, the viability of *A. suum* eggs in the controls (G-SAs) was reduced by 7–13%. When eggs were placed in areas containing antiparasitic fungi (G-2CAs), there was a decrease in the viable eggs of the nematode by 17%, as recorded after 15 days ($F = 23.125$, $p = 0.037$), and 64.9% on day 30 ($F = 20.892$, $p = 0.036$).

By considering data acquired in the respective control and treated groups, the percentages of reduction were normalized (Table 1).

Table 1. Adjusted reduction (%) of viable *A. suum* eggs exposed to the antiparasitic fungi *M. circinelloides* and *D. flagrans*.

Day	Trays	Small Ground Areas
15	13.1	10.6
30	58.3	51.7

3.2. Effect on Cyathostomin Third-Stage Larvae (L3s)

As given in Figure 6, no larvae were detected at the beginning of the study. The percentages of L3 developed in the control trays were 20% on day 15 and 60.4% on day 30. In the trays with fungi, the percentages of larvae recovered were 10.7% and 16.5%, respectively.

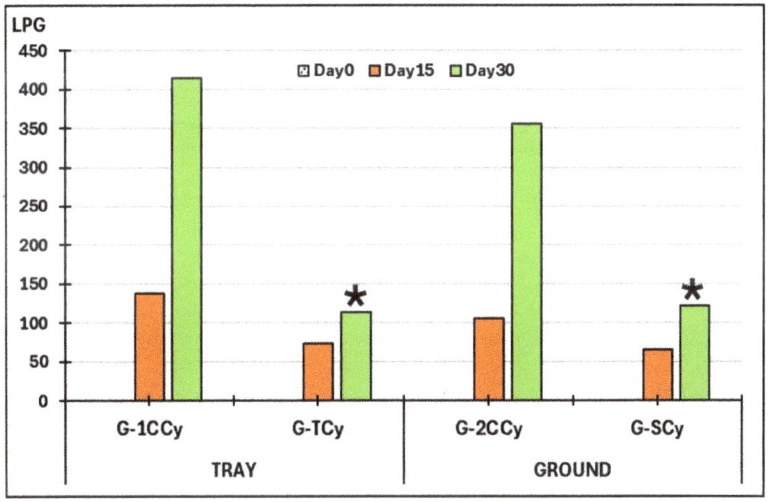

Figure 6. Variations on the viability of L3 larvae of cyathostomins. G-1CCy: control group trays (seeds without fungi); G-TCy: trays sown with seeds soaked in submerged culture containing *M. circinelloides* and *D. flagrans*: G-2CCy: control soil parcels (without fungi); G-SCy: soil parcels sown with seeds soaked in submerged culture containing *M. circinelloides* and *D. flagrans*. LPG: L3 larvae of cyathostomins per gram of feces; (*): significant differences.

Regarding the small ground areas serving as controls, a percentage of 15.3% and 52% L3 larvae evolved from cyathostomin eggs on days 15 and 30, respectively. Concerning the exposure to antiparasitic fungi, the percentages recorded were 9.5% and 17.7% on days 15 and 30. Significant differences were recorded at the end of the study in the trays ($F = 22.381$, $p = 0.031$) and in the ground ($F = 18.231$, $p = 0.042$).

Table 2 shows that the PR_{L3} in the presence of fungi was higher in the trays with herbage than in small ground areas, but these differences were not significant ($p > 0.05$).

Table 2. Reduction (%) of cyathostomin L3 larvae exposed to the antiparasitic fungi *M. circinelloides* and *D. flagrans*.

Day	Trays	Small Ground Areas
15	46.7	38.1
30	72.7	66.0

4. Discussion

Captive wild animals in zoos are at high risk of infection by parasites that develop some stages in the soil, and the problem increases when the animals are kept in plots with vegetation favoring certain parasites with a direct life cycle (as is the case with numerous helminths) survive and complete their evolvement in the soil [17,18]. Consequently, an investigation was designed to try to reduce the numbers and survival of infective parasitic stages in the soil by ensuring the presence of two antiparasitic fungi. For this purpose, forage seeds were soaked in a mixture of saprophytic filamentous fungi (*M. circinelloides* and *D. flagrans*) before sowing in plastic trays. Once the seeds germinated, feces of piglets shedding eggs of the gastrointestinal nematode *Ascaris suum* and feces of horses passing eggs of cyathostomins were placed. After a 30-day period, the counts of *A. suum* viable eggs reduced significantly by two-thirds, and those of cyathostomin third-stage (L3) larvae by three-quarters. These results are partially in agreement with previous studies involving fungal species such as *M. circinelloides* [19], *Trichoderma atrobrunneum* F.B. Rocha, Chaverri & Jaklitsch [20], or *Clonostachys rosea* Schroers, Samuels, Seifert & W. Gams [21] directly sprayed on the feces of pigs [22,23], where similar reduction percentages of two-thirds for *Ascaris suum* and one-third for strongylid nematodes were found.

Parasite control among wild animals kept in a zoo basically consists of administering drugs designed for livestock because neither the proper dosage nor the possible adverse effects are well known [5,7,12], and specific indications cannot be found in the commercial leaflets. Besides this, deworming does not affect the infective stages in the soil, and the risk of infection maintains and even increases as the pasture rotation practices suggested for livestock are not applicable in most zoos [24]. Because many infections provoked by gastrointestinal parasites originate when infective stages developing in the soil are ingested [25], i.e., while grazing, the need to have a realistic procedure to limit the numbers of these infective stages seems essential. Proper spreading of spores of soil filamentous fungi has been extensively tried, encompassing mostly their oral administration through different edible formulations, with the aim of the fungal spores are passed in the feces together with some parasitic stages, and once in the soil, develop their mycelia and act against the pathogens [26]. By means of water solutions and milled cereals [13], the spores of several antiparasitic fungi were distributed among captive equids in a zoological garden. Later, and based on the results achieved in horses maintained under continuous pasturing, nutritional pellets elaborated with chlamydospores of *M. circinelloides* and *D. flagrans* in an industrial factory provided highly successful results among different wild herbivores infected by strongylid nematodes [27]. In the current investigation, the usefulness of soaking forage seeds with a submerged culture containing a blend of *M. circinelloides* and *D. flagrans* before sowing them in the ground has been demonstrated by observing that the counts of the pig roundworm *A. suum* eggs and horse cyathostomins were reduced by 51.7% and 66%, respectively, after a period of 30 days. So far, this is the first and only contribution to prevent infection among captive animals and to limit the application of chemical dewormers without oral administration because the risk of infection can be significantly reduced with this strategy (by half and two-thirds, respectively). Another interesting point lies in the fact that no additional tasks are required for animal keepers, and the antagonistic and beneficial effect is achieved without depending on animal intake, palatability, etc. In a recent investigation, the administration of edible gelatins made with chlamydospores of *M. circinelloides* and *D. flagrans* reduced the viability of the gastrointestinal nematode *Trichuris* sp. in the feces of captive baboons by 44% [11], no evidence regarding the influence of soil types or extreme weather conditions on the development of filamentous fungi are available. It is well known that under adverse conditions, fungi remain latent as spores.

During the last two decades, many investigations have analyzed the practical possibilities of using parasitophagous fungi for biological control among animals and even people [17], with an important stress on the different and practical ways which ensure their proper distribution. Although the oral administration of parasiticidal fungal spores has provided promising results, its practical application entails certain difficulties, mainly consisting of the fact that wild animals might refuse to take feed containing the fungi. In the current assay, the usefulness of a totally novel application has been tested. The novelty consisted of soaking seeds of forage species in a submerged culture containing *M. circinelloides* and *D. flagrans* prior to seeding a small land area, with the aim of contributing to the spreading of that fungus along the land. Accordingly, the results obtained in the present investigation could be directly extrapolated to livestock reared on farms. Although it appears very interesting, the possibility of applying the proposed strategy on wildlife reserves appears very difficult due to the problems associated with spreading the fungal spores.

Ascarids and strongylid nematodes (such as cyathostomins) are parasites frequently identified mainly in the feces of animals that enjoy plots with soil and/or vegetation, and their presence persists despite the regular administration of effective anthelmintics [5,8,9,18,24]. A remarkable result in the present investigation was the observation of a reduction in the number of viable *A. suum* eggs by 10.1% at fifteen days, in contrast to 38.1% with respect to L3 cyathostomins. It should be taken into account that the roundworm eggshell provides protection against adverse environmental conditions (freezing, dryness, UV radiation...), which would explain the need for a certain amount of time to elapse before fungi are able to damage it, break it, and penetrate inside it [10,22,23,28].

Finally, the tests carried out in trays proved to be a very suitable procedure for the germination of forage species and for the distribution of chlamydospores and the development of mycelia belonging to the parasiticidal fungi *M. circinelloides* and *D. flagrans*, which is concluded considering the reduction percentages recorded in the numbers of viable eggs of *A. suum*, as well as in the counts of L3 cyathostomins.

5. Conclusions

Controlling parasites that are developing their infective stages in areas with land and vegetation requires preventive measures to reduce the risk of infection. Sowing the ground with seeds previously soaked in a submerged medium for culturing the parasitophagous fungi *Mucor circinelloides* and *Duddingtonia flagrans* offers an innovative and effective strategy to lessen the numbers of viable *A. suum* viable eggs and the counts of L3 cyathostomin. The performance of tests in trays for assessing antiparasitic offers a model that can be extrapolated to those developed in soil, so they are highly recommended.

Author Contributions: Conceptualization, J.Á.H., C.C.-M., A.P.-S., R.M., M.S.A. and M.C.; methodology, C.V., J.L., I.Z., M.B., I.A.-R., R.S., E.V. and G.P.-A.; software, J.Á.H. and M.C.; validation, C.C.-M., R.M. and M.S.A.; formal analysis, J.L., R.S., G.P.-A., M.S.A. and M.C.; investigation, C.V., M.B. and I.A.-R.; resources, E.V. and C.C.-M.; data curation, J.L., I.A.-R., G.P.-A. and R.M.; writing—original draft preparation, J.Á.H., C.V., I.A.-R. and R.S.; writing—review and editing, J.Á.H., A.P.-S., R.M., J.L. and M.S.A.; visualization, M.B., I.A.-R. and R.S.; supervision, J.Á.H., R.M. and G.P.-A.; project administration, I.Z., C.C.-M. and M.S.A.; funding acquisition, J.Á.H., A.P.-S., M.S.A. and M.C. All authors have read and agreed to the published version of the manuscript.

Funding: This trial was partly supported by the Ministerio de Ciencia e Innovación (Spain) and the European Regional Development Fund (ERDF) (Research Project PID2020-120208RB-I00), the Consellería de Cultura, Educación e Universidades (Xunta de Galicia, Spain) (Research Project ED431B2021/07) and the Consellería do Medio Rural (Xunta de Galicia, Spain) (Research Project AEI 2022 Operational Groups).

Institutional Review Board Statement: Not applicable.

Data Availability Statement: Dataset available on request from the authors.

Acknowledgments: The authors express their sincere gratitude to the Head of the "Granja Gayoso Castro" (Deputación Provincial de Lugo, Spain) and to "Marcelle Natureza Zoological Park" for their valuable collaboration.

Conflicts of Interest: The authors declare no conflicts of interest. The funders had no role in the design of the study; in the collection, analyses, or interpretation of data; in the writing of the manuscript; or in the decision to publish the results.

References

1. Rabb, G.B. The Evolution of Zoos from Menageries to Centers of Conservation and Caring. *Curator. Mus. J.* **2004**, *47*, 237–246. [CrossRef]
2. Packer, J.; Ballantyne, R. The Role of Zoos and Aquariums in Education for a Sustainable Future. *New Dir. Adult Contin. Educ.* **2010**, *2010*, 25–34. [CrossRef]
3. Conway, W.G. Buying Time for Wild Animals with Zoos. *Zoo Biol.* **2011**, *30*, 1–8. [CrossRef]
4. Mellen, J.D. Zoo Standards for Keeping Small Felids in Captivity. Available online: https://nagonline.net/763/zoo-standards-keeping-small-felids-captivity/ (accessed on 13 February 2024).
5. Gonzálvez, M.; Moreno, E.; Pérez-Cutillas, P.; Gilbert, T.; Ortiz, J.; Valera, F.; Espeso, G.; Benzal, J.; Ibáñez, B.; de Ybáñez, M.d.R.R. Zoological Institutions as Hotspots of Gastrointestinal Parasites That May Affect the Success of Ungulate Reintroduction Programmes. *Vet. Rec.* **2021**, *189*, e506. [CrossRef]
6. Capasso, M.; Maurelli, M.P.; Ianniello, D.; Alves, L.C.; Amadesi, A.; Laricchiuta, P.; Silvestre, P.; Campolo, M.; Cringoli, G.; Rinaldi, L. Use of Mini-FLOTAC and Fill-FLOTAC for Rapidly Diagnosing Parasitic Infections in Zoo Mammals. *Rev. Bras. Parasitol. Vet.* **2019**, *28*, 168–171. [CrossRef] [PubMed]
7. Garijo, M.M.; Ortiz, J.M.; Ruiz de Ibáñez, M.R. Helminths in a Giraffe (Giraffa Camelopardalis Xgiraffa) from a Zoo in Spain. *Onderstepoort J. Vet. Res.* **2004**, *71*, 153–156. [CrossRef]
8. Atanaskova, E.; Kochevski, Z.; Stefanovska, J.; Nikolovski, G. Endoparasites in Wild Animals at the Zoological Garden in Skopje, Macedonia. *J. Threat. Taxa* **2011**, *3*, 1955–1958. [CrossRef]
9. Nath, T.C.; Eom, K.S.; Choe, S.; Hm, S.; Islam, S.; Ndosi, B.A.; Kang, Y.; Bia, M.M.; Kim, S.; Eamudomkarn, C.; et al. Insight into One Health Approach: Endoparasite Infections in Captive Wildlife in Bangladesh. *Pathogens* **2021**, *10*, 250. [CrossRef] [PubMed]
10. Hernández, J.A.; Cazapal-Monteiro, C.F.; Arroyo, F.L.; Silva, M.I.; Palomero, A.M.; Paz-Silva, A.; Sánchez-Andrade, R.; Arias, M.S. Biological Control of Soil Transmitted Helminths (STHs) in a Zoological Park by Using Saprophytic Fungi. *Biol. Control* **2018**, *122*, 24–30. [CrossRef]
11. Paz-Silva, A.; Salmo, R.; Viña, C.; Miguel Palomero, A.; Ángel Hernández, J.; Sánchez-Andrade, R.; Cazapal-Monteiro, C.; Sol Arias, M. Gelatin Treats Containing Filamentous Fungi to Promote Sustainable Control of Helminths among Pets and Zoo Animals. *Biol. Control* **2023**, *179*, 105184. [CrossRef]
12. Salmo, R.; Viña, C.; Lozano, J.; Palomero, A.M.; Hernández, J.Á.; Bonilla, R.; Sánchez-Andrade, R.; Paz-Silva, A.; Madeira de Carvalho, L.M.; Arias, M.S.; et al. Saprophytic Filamentous Fungi against Helminths Affecting Captive Wild Animals. *Encyclopedia* **2024**, *4*, 91–100. [CrossRef]
13. Arias, M.S.; Cazapal-Monteiro, C.F.; Suárez, J.; Miguélez, S.; Francisco, I.; Arroyo, F.L.; Suárez, J.L.; Paz-Silva, A.; Sánchez-Andrade, R.; Mendoza de Gives, P. Mixed Production of Filamentous Fungal Spores for Preventing Soil-Transmitted Helminth Zoonoses: A Preliminary Analysis. *BioMed Res. Int.* **2013**, *2013*, 567876. [CrossRef]
14. Hernández, J.A.; Vázquez-Ruiz, R.A.; Cazapal-Monteiro, C.F.; Valderrábano, E.; Arroyo, F.L.; Francisco, I.; Miguélez, S.; Sánchez-Andrade, R.; Paz-Silva, A.; Arias, M.S. Isolation of Ovicidal Fungi from Fecal Samples of Captive Animals Maintained in a Zoological Park. *J. Fungi* **2017**, *3*, 29. [CrossRef] [PubMed]
15. Cernea, M.; Madeira de Carvalho, L.; Vasile, C. *Atlas of Diagnosis of Equine Strongylidosis*; Editura Academic Press: Târgovişte, Romania, 2008; ISBN 978-973-744-127-0.
16. Araújo, J.V.d.; Fonseca, J.d.S.; Barbosa, B.B.; Valverde, H.A.; Santos, H.A.; Braga, F.R. The Role of Helminthophagous Fungi in the Biological Control of Human and Zoonotic Intestinal Helminths. *Pathogens* **2024**, *13*, 741. [CrossRef] [PubMed]
17. VanHoy, G. Overview of Gastrointestinal Parasites of Ruminants. Available online: https://www.msdvetmanual.com/digestive-system/gastrointestinal-parasites-of-ruminants/overview-of-gastrointestinal-parasites-of-ruminants#Prevention_v81479076 (accessed on 18 August 2024).
18. Esteban-Sánchez, L.; García-Rodríguez, J.J.; García-García, J.; Martínez-Nevado, E.; de la Riva-Fraga, M.A.; Ponce-Gordo, F. Wild Animals in Captivity: An Analysis of Parasite Biodiversity and Transmission among Animals at Two Zoological Institutions with Different Typologies. *Animals* **2024**, *14*, 813. [CrossRef] [PubMed]
19. Van Tieghem, P. Nouvelles recherches sur les Mucorinées. *Fortin Masson* **1875**, *6*, 175.
20. Chaverri, P.; Branco-Rocha, F.; Jaklitsch, W.; Gazis, R.; Degenkolb, T.; Samuels, G.J. Systematics of the *Trichoderma Harzianum* Species Complex and the Re-Identification of Commercial Biocontrol Strains. *Mycologia* **2015**, *107*, 558–590. [CrossRef]
21. Schroers, H.-J.; Samuels, G.J.; Seifert, K.A.; Gams, W. Classification of the Mycoparasite *Gliocladium Roseum* in *Clonostachys* as *C. Rosea*, Its Relationship to *Bionectria Ochroleuca*, and Notes on Other *Gliocladium*-like Fungi. *Mycologia* **1999**, *91*, 365–385. [CrossRef]
22. Cortiñas, F.J.; Cazapal-Monteiro, C.F.; Hernández, J.A.; Arroyo, F.L.; Miguélez, S.; Suárez, J.; López de Arellano, M.E.; Sánchez-Andrade, R.; Mendoza de Gives, P.; Paz-Silva, A.; et al. Potential Use of *Mucor Circinelloides* for the Biological Control of Certain Helminths Affecting Livestock Reared in a Care Farm. *Biocontrol Sci. Technol.* **2015**, *25*, 1443–1452. [CrossRef]

23. Viña, C.; Silva, M.I.; Palomero, A.M.; Voinot, M.; Vilá, M.; Hernández, J.Á.; Paz-Silva, A.; Sánchez-Andrade, R.; Cazapal-Monteiro, C.F.; Arias, M.S. The Control of Zoonotic Soil-Transmitted Helminthoses Using Saprophytic Fungi. *Pathogens* **2020**, *9*, 1071. [CrossRef] [PubMed]
24. Rapiya, M.; Hawkins, H.-J.; Muchenje, V.; Mupangwa, J.F.; Marufu, M.C.; Dzama, K.; Mapiye, C. Rotational Grazing Approaches Reduces External and Internal Parasite Loads in Cattle. *Afr. J. Range Forage Sci.* **2019**, *36*, 151–159.
25. Mendoza-de Gives, P. Soil-Borne Nematodes: Impact in Agriculture and Livestock and Sustainable Strategies of Prevention and Control with Special Reference to the Use of Nematode Natural Enemies. *Pathogens* **2022**, *11*, 640. [CrossRef]
26. Hernández Malagón, J.Á.; Filipa Cazapal-Monteiro, C.; Bonilla Quintero, R.; Miguel Palomero Salinero, A.; Isabel Silva Torres, M.; Voinot Messnier, M.; Vilá Pena, M.; Romasanta Blanco, Á.; Pedreira García, J.; Paz Silva, A.; et al. Advantageous Fungi against Parasites Transmitted through Soil. In *Fungal Infection*; Silva de Loreto, É., Simoni Moraes Tondolo, J., Eds.; IntechOpen: London, UK, 2019; ISBN 978-1-83880-468-8.
27. Palomero, A.M.; Cazapal-Monteiro, C.F.; Valderrábano, E.; Paz-Silva, A.; Sánchez-Andrade, R.; Arias, M.S. Soil Fungi Enable the Control of Gastrointestinal Nematodes in Wild Bovidae Captive in a Zoological Park: A 4-Year Trial. *Parasitology* **2020**, *147*, 791–798. [CrossRef] [PubMed]
28. Cruz, L.M.; Allanson, M.; Kwa, B.; Azizan, A.; Izurieta, R. Morphological Changes of Ascaris Spp. Eggs During Their Development Outside the Host. *J. Parasitol.* **2012**, *98*, 63–68. [CrossRef]

Journal of
Zoological and Botanical Gardens

MDPI

Article

The Difficulties of Ex Situ Conservation: A Nationwide Investigation of Avian Haemosporidia Among Captive Penguins in Japan

Mizue Inumaru [1,2,†], Misa Shimizu [2,3,†], Ayumi Shibata [2,†], Koichi Murata [4] and Yukita Sato [2,5,*]

1 Department of Medical Entomology, National Institute of Infectious Diseases, Toyama 1-23-1, Shinjuku 162-8640, Japan; inumaru-m@niid.go.jp
2 Laboratory of Biomedical Science, Department of Veterinary Medicine, College of Bioresource Sciences, Nihon University, 1866 Kameino, Fujisawa 252-0880, Japan; misa.shimizu@gamtc.lt (M.S.); hmatc10559300vet11@gmail.com (A.S.)
3 P. B. Sivickis Laboratory of Parasitology, Nature Research Centre, Akademijos 2, 08412 Vilnius, Lithuania
4 Yokohama Zoological Gardens "ZOORASIA", 1175-1 Kamishiranecho, Asahiku, Yokohama 241-0001, Japan; haemoproteus@gmail.com
5 Laboratory of Veterinary Parasitology, Department of Veterinary Medicine, Faculty of Agriculture, Iwate University, 3-18-8 Ueda, Morioka 020-8550, Japan
* Correspondence: yukita@iwate-u.ac.jp
† These authors contributed equally to this work.

Abstract: Avian malaria has been a continuous problem in both wild and captive populations of penguins throughout the world. In Japan, where there are over 3000 captive penguins, avian malaria (by *Plasmodium* spp.) and haemoproteosis (by *Haemoproteus* spp.) have been sporadically detected throughout the country. However, no comprehensive studies have been carried out, and the national status of infection has been unknown until now. In this study, the prevalence and lineage composition of haemosporidian parasites was investigated in captive penguins throughout Japan for the first time. A total of 1203 penguins from 55 facilities were sampled from January 2010 to December 2019. Parasites were detected by nested PCR and microscopy of blood and tissue samples. The total prevalence was 7.48% for *Plasmodium* and 1.75% for *Haemoproteus*, of which some are suggested to have been acquired during the study period. The odds of infection were higher in individuals kept outdoors compared to indoors, re-confirming that exposure to vectors is one of the major factors. Additionally, the odds of death were higher in infected individuals, although differences between parasite lineages were also observed. This study provides an overview of avian malaria in penguins of Japan in the hope of guiding future studies and conservation actions in captivity.

Keywords: avian malaria; *Haemoproteus*; haemosporidia; Japan; national survey; penguin; *Plasmodium*; prevalence

1. Introduction

Avian malaria is a vector-borne disease caused by *Plasmodium* parasites. Along with closely related haemosporidian parasites of the genus *Haemoproteus* and *Leucocytozoon*, these parasites are distributed throughout the world [1]. These avian haemosporidian parasites can cause varying levels of pathogenicity in host birds depending on the parasite species and host species. Infection can sometimes be extremely pathogenic for some avian hosts, especially if the host species have historically had little to no exposure to these pathogens [1–3]. In Hawaii, numerous endemic honeycreepers went extinct due to the introduction of avian malaria and avian poxvirus, which are both pathogens that had not been native in Hawaiian islands historically [2,4,5]. Similarly, cases of death due to avian haemosporidia have been reported from many captive birds outside of their natural distribution, including cranes [6], parrots [7,8], passerine birds [9], owls [10,11], and penguins [12,13].

Many species of penguins are currently endangered due to human-caused threats such as pollution, climate change, habitat loss, and infectious diseases [14]. To conserve these endangered birds, populations in captivity have been established throughout the world. While species of the Antarctic or sub-Antarctic region are generally kept indoors, temperate species are often kept in outdoor enclosures where vector insects such as mosquitoes and biting midges can freely access the birds [15]. These penguins are therefore constantly exposed to vectors that can transmit avian haemosporidia from wild birds of surrounding areas [1,16–18]. As most penguin species naturally inhabit areas free of avian haemosporidia or areas with low prevalence, they are extremely sensitive to these pathogens. Consequently, many fatal cases due to avian haemosporidia, particularly avian malaria, have been reported [12,16,19–21]. Avian haemosporidia are now considered one of the most significant pathogens in the conservation of penguins [12,22–24].

In Japan, there are over 3000 penguins in zoos and aquariums across the nation [25,26]. The first report of avian haemosporidia in captive penguins of Japan was from Magellanic penguins (*Spheniscus magellanicus*) of Mie Prefecture in 1987 [27]. Since then, avian malaria (*Plasmodium* spp.) and haemoproteosis (*Haemoproteus* spp.) have been sporadically detected throughout Japan [19,20,28,29]. Furthermore, avian malaria has been detected from *Culex pipiens* group mosquitoes caught at zoological facilities, including individuals that had fed on penguins and wild birds [17,29,30]. These reports strongly suggest that avian haemosporidia are actively transmitted at such facilities in Japan between captive penguins and wild birds that inhabit the surrounding areas. However, the prevalence and lineage composition, which are important for considering conservation actions, of avian haemosporidia in penguins across Japan were unknown. The prevalence and lineage composition of haemosporidian parasites in captive penguins throughout the country were investigated in this study. The environment of the enclosure and vital status were incorporated in the analysis to reveal possible risk factors for ex situ populations of endangered penguins.

2. Materials and Methods

2.1. Sample Collection

Whole blood from captive penguins was sampled between January 2010 and December 2019 at 55 facilities of 29 prefectures across Japan. Note that individuals from two previously reported studies are included as part of this study [28,29]. The environment in which these penguins were kept differed by species and facility. Each individual was classified into two environmental groups: outdoor enclosure (outdoor group) and indoor enclosure (indoor group). The indoor group included individuals that were maintained indoors but allowed outdoors during the winter months (November to April). All deceased individuals belonged to the outdoor group. Two Humboldt penguins (*Spheniscus humboldti*) passed away during the study period and were included in the deceased group. Although information including relocation history between facilities, prophylactic medication history, symptoms and treatments during sampling were provided by some facilities, not all individuals were provided with this information.

In live penguins, blood was obtained from either the brachial or metatarsal vein. Some individuals were sampled multiple times mainly for routine medical check-ups, although the frequency and span of sampling differed between individuals and facilities. In penguins that died, a small portion of organ tissue such as liver, spleen and lung was obtained. All samples were kept in microtubes with 70% ethanol. Blood smears were prepared when possible.

All samples and blood smears were then sent to the Laboratory of Biomedical Science of Nihon University College of Bioresource Sciences. The blood and tissue samples were kept at −20 °C until DNA isolation. Blood smears were fixed with 100% methanol and stained with Hemacolor® (Merck KGaA, Darmstadt, Germany). Each blood smear was mounted with a cover glass using a mounting medium (O. Kindler GmbH, Freiburg, Germany) after confirming that they were dry.

All procedures for collecting samples from birds in this study were performed in accordance with the ethical standards of the Act on Welfare and Management of Animals 1973.

2.2. DNA Extraction and Molecular Detection of Avian Haemosporidia

DNA was extracted from the blood or tissue samples using either the standard phenol–chloroform method or QIAamp®DNA Micro Kit (QIAGEN, Hilden, Germany). Tris-EDTA buffer was used as the final buffer to dissolve the extracted DNA. DNA concentration was confirmed and adjusted to 50 ng/μL using Nanodrop One Microvolume UV-Vis Sprectrophotometers (Thermo Fisher Scientific, Waltham, MA, USA). Then, a nested polymerase chain reaction (PCR) targeting the partial mitochondrial cytochrome *b* (cyt*b*) gene of avian haemosporidia was carried out using a previously described protocol [29]. In brief, HaemNFI/HaemNR3 primer set was used for the 1st PCR. For the 2nd PCR, the HaemF/HaemR2 primer set was used for *Plasmodium* spp. and *Haemoproteus* spp., and the HaemFL/HaemR2L primer set was used for *Leucocytozoon* spp. [31]. The PCR products were then visualized using 1.5% agarose gels (Agarose S: Nippon Gene, Chiyoda, Japan) containing ethidium bromide (Nacalai tesque, Nakagyo, Japan). All obtained nucleotide sequences were compared with sequences in the GenBank database using the Basic Local Alignment Search Tool [32,33] and sequences in the MalAvi database [34].

2.3. Microscopic Detection and Infection Intensity of Avian Haemosporidia

Blood smears were examined under an Olympus BX43 of Olympus IX71 light microscope (Olympus, Tokyo, Japan). The smears were screened at 400× magnification and then carefully examined at 1000× magnification under oil immersion if a suspected creature was observed. Photos of the observed parasites were taken with cellSens Standard 1.6 (Olympus, Tokyo, Japan) and then morphologically classified into the proper genera [1]. Afterwards, intensity of infection (parasitemia) was estimated by counting the number of parasites per 10,000 erythrocytes. Counting of the erythrocytes was started at a random location, although repositioned in cases where there were overlapping erythrocytes. When an infected erythrocyte was found outside of the counted 10,000 erythrocytes, the parasitemia was calculated as 1/100,000 erythrocytes.

2.4. Statistical Analysis

To test whether infections were acquired over time, the initial and total parasite prevalence was compared using Fisher's exact test. The initial prevalence was calculated by the infection status of the first sample for each individual. The total prevalence was calculated by the infection status of all samples per individual. An individual was considered positive if one or more samples were positive by PCR and negative if all samples were negative by PCR. Among only individuals for which morphological detections were possible, the prevalence by PCR and by microscopy was compared using Fisher's exact test.

Parasite prevalence was compared with Fisher's exact test between environmental groups and vital status (live and deceased individuals). To reduce biases due to repeated sampling of particular individuals, only the first sample for each individual was used for the above analysis. The parasitemia and prevalence of each parasite lineage were compared between vital status using Student's *t*-test and Fisher's exact test, respectively.

All statistical analyses were carried out with R version 4.3.2 [35] and were adjusted with Bonferroni correction. The 5% significance level was used to determine statistical significance throughout the study.

3. Results

3.1. Haemosporidian Prevalence by PCR

A total of 1966 samples from 1203 individuals of 12 species were collected. Collectively, 110 individuals of seven species in 18 prefectures were positive by PCR for either *Plasmodium* or *Haemoproteus* (Table 1, Supplementary Table S1). Note that one southern rockhopper penguin (*Eudyptes chrysocome*) was co-infected with *Plasmodium* and *Haemoproteus* parasites. *Leucocytozoon* was not detected in this study. Positive individuals were detected in all eight areas of Japan (Table S1). The total prevalence was 7.48% for *Plasmodium* and 1.75% for *Haemoproteus*. The initial prevalence using only the first sample of each individual was 5.40% and 1.08% for *Plasmodium* and *Haemoproteus*, respectively. There was no significant difference in prevalence between initial and total prevalence for *Haemoproteus* (Fisher's exact test: $p = 0.22$), even when compared among only live individuals of the outdoor group (Fisher's exact test: $p = 0.19$) and among only individuals that were sampled multiple times (Fisher's exact test: $p = 0.07$). Meanwhile, for *Plasmodium*, the total prevalence was significantly higher than the initial prevalence in all cases (Fisher's exact test: all samples $p = 0.05$; only live individuals of outdoor group $p = 0.03$; only multiple-sampled individuals $p < 0.01$).

Among individuals of the indoor group, three and one individual(s) were positive for *Plasmodium* and *Haemoproteus*, respectively. It is important to note that all four positive individuals had experienced relocation from a different facility where they were kept in outdoor enclosures. All other positive individuals were those of the outdoor group. The *Plasmodium* prevalence and odds of infection were significantly higher in outdoor individuals compared to indoor individuals (Fisher's exact test: OR = 6.37 $p < 0.01$), even when only live individuals were included (Fisher's exact test: OR = 4.93 $p < 0.01$; Figure 1). Meanwhile, although not statistically significant, the odds of infection were higher in outdoor individuals for *Haemoproteus* as well, both with and without deceased individuals (Fisher's exact test: all individuals OR = 3.52 $p = 0.32$, only live individuals OR = 3.42 $p = 0.31$; Figure 1).

By vital status, *Plasmodium* prevalence was 30.61% for deceased individuals and 5.32% for live individuals. The prevalence was significantly higher for deceased individuals compared to live individuals, even when compared among only outdoor individuals (Fisher's exact test: all individuals OR 9.95 $p < 0.01$, only outdoor individuals OR = 8.02 $p < 0.01$; Figure 1). Meanwhile, there was no significant difference between environmental groups for *Haemoproteus*, both with and without deceased individuals (Fisher's exact test: all individuals $p = 0.44$ odds ratio 1.86, only live individuals OR 1.55 $p = 0.50$; Figure 1).

Table 1. Summary of haemosporidian detection in penguins of this study, by species, environmental factor, and vital status.

Parasite Genus	Species [a]	Live						Deceased (All Outdoor)			Total		
		Indoor (Complete and Partial)			Outdoor								
		Individuals	Positive	Prevalence	Individuals	Positive [b]	Prevalence [b]	Individuals	Positive [b]	Prevalence [b]	Individuals	Positive [b]	Prevalence [b]
Plasmodium	King penguin (*Ap. patagonicus*)	36	0	0	18	0	0	3	0	0	57	0	0
	Emperor penguin (*Ap. forsteri*)	2	0	0							2	0	0
	Adelie penguin (*Py. adeliae*)	14	0	0							14	0	0
	Chinstrap penguin (*Py. antarcticus*)	15	0	0							15	0	0
	Gentoo penguin (*Py. papua*)	69	0	0	10	0 (1)	0 (10.0)				79	0 (1)	0 (1.27)
	Fairy penguin (*El. minor*)				2	0	0	10	0	0	12	0	0
	Humboldt penguin (*Sp. humboldti*)	21	0	0	574	27 (34)	4.70 (5.92)	23	12 (14)	52.17 (60.87)	618	39 (48)	6.31 (7.77)
	Magellanic penguin (*Sp. magellanicus*)	58	1	1.72	88	14 (19)	15.91 (21.59)	4	2	50.00	150	17 (22)	11.33 (14.67)
	African penguin (*Sp. demersus*)	24	1	4.17	167	4 (14)	2.40 (8.38)	7	0	0	198	5 (15)	2.52 (7.58)
	Macaroni penguin (*Es. chrysolophus*)	1	1	100							1	1	100
	N. rockhopper penguin (*Es. moseleyi*)	23	0	0	7	0	0				30	0	0
	S. rockhopper penguin (*Es. chrysocome*)	8	0	0	14	1	7.14	5	2	40.00	27	3	11.11
	total	271	3	1.11	880	46 (69)	5.23 (7.84)	52	16 (18)	30.77 (34.62)	1203	65 (90)	5.40 (7.48)
Haemoproteus	King penguin (*Ap. patagonicus*)	36	0	0	18	0	0	3	0	0	57	0	0
	Emperor penguin (*Ap. forsteri*)	2	0	0							2	0	0
	Adelie penguin (*Py. adeliae*)	14	0	0							14	0	0
	Chinstrap penguin (*Py. antarcticus*)	15	0	0							15	0	0
	Gentoo penguin (*Py. papua*)	69	0	0	10	1	10.00				79	1	1.27
	Fairy penguin (*El. minor*)				2	0	0	10	0	0	12	0	0
	Humboldt penguin (*Sp. humboldti*)	21	0	0	574	2 (5)		23	1	4.35	618	3 (6)	0.49 (0.97)
	Magellanic penguin (*Sp. magellanicus*)	58	0	0	88	4 (7)	4.55 (7.95)	4	0	0	150	4 (7)	2.67 (4.67)
	African penguin (*Sp. demersus*)	24	1	4.17	167	4 (5)	2.40 (3.00)	7	0	0	198	5 (6)	2.52 (3.03)
	Macaroni penguin (*Es. chrysolophus*)	1	0	0							1	0	0
	N. rockhopper penguin (*Es. moseleyi*)	23	0	0	7	0	0				30	0	0
	S. rockhopper penguin (*Es. chrysocome*)	8	0	0	14	0 (1)	0 (7.14)	5	0	0	27	0 (1)	0 (3.70)
	total	271	1	0.37	880	11 (19)	1.25 (2.16)	52	1	1.92	1203	13 (21)	1.08 (1.75)

[a] Genera are abbreviated: *Ap.* = *Aptenodytes*, *Py.* = *Pygoscelis*, *El.* = *Eudyptula*, *Sp.* = *Spheniscus*, *Es.* = *Eudyptes*. [b] Numbers in parentheses show the total number of positive individuals and prevalence (i.e., parasites detected at least once). Numbers in front of parentheses show the initial number of positive individuals and prevalence (i.e., infection status of only the first sample of each individual).

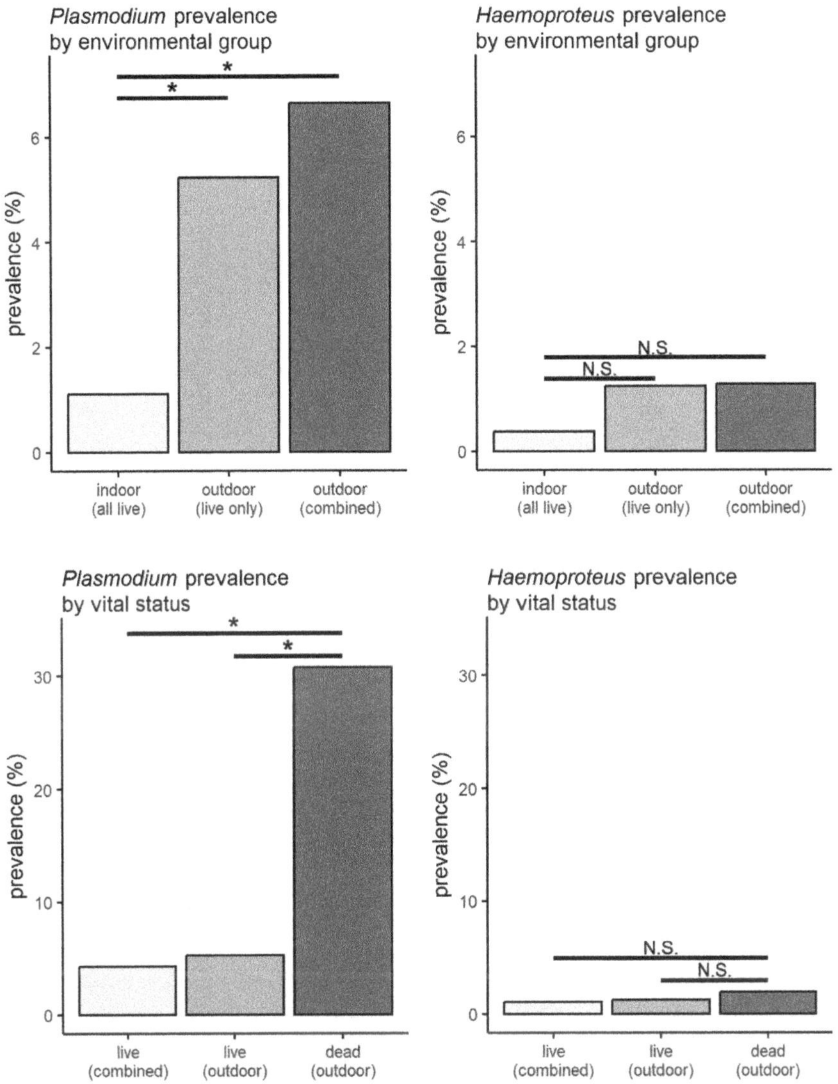

Figure 1. Parasite prevalences of penguins in this study, by environmental group and vital status. Asterisks (*) show significant differences ($p < 0.05$). N.S. = not significant.

3.2. Phylogeny

In total, 13 *Plasmodium* lineages and three *Haemoproteus* lineages were detected in this study (Table 2, Figure 2). Two *Haemoproteus* lineages named SPHUM06 and PYGPAP01 were added to the GenBank (accession nos. LC853581 and LC853582) and MalAvi database [34]. Lineages CXINA01, CXPIP10, GRW06, SPHUM03, SPHUM05, SYCON02, and TURPAL01 were detected from penguins for the first time in Japan (note that SPHUM03 and SPHUM05, which were reported in our previous studies, are included as part of this study). Combined with previously detected lineages, 27 parasite lineages were detected from penguins of Japan (Table 2).

Table 2. Summary of *Plasmodium* and *Haemoproteus* lineage that have been detected in penguins of Japan, based on MalAvi and GenBank.

Lineages [a]	Distribution		Species [b]	Detected Species in Japan [c]	Reference
pAPTPAT01	Japan only	host	1	***Ap. patagonicus* (HK, TY)**	MalAvi [d]
pCXINA01 *	Japan only	host	2	***Sp. humbodti* (TY, KU)**; ***Es. chrysocome* (TY)**	this study
		vector	1	*Cx. inatomii* (NI)	[36]
pCXPIP09 *	Japan only	host	11	*Larus argentatus* (CB); ***El. minor* (TY)**; ***Sp. humboldti* (HK, AK, TY, NI, KN, OS)**; ***Sp. magellanicus* (TY, NI, YN, HG, OI)**; ***Sp. demersus* (TY, KY, FO)**; ***Es. chrysolophus* (TY)**; ***Es. chrysocome* (NI)**; *Ardea cinirea* (NS); *Cyanopica cyanus* (TY); *Corvus corone* (CB); *Corvus macrorhynchos* (TY)	[19,29,37,38]; MalAvi; this study
		vector	5	*Cx. pipiens* (TY, KN); *Cx. inatomii* (NI); *Cx. sasai* (TY); *Lt. vorax* (KN)	[17,36,39,40]
pCXPIP10 *	Asia, Europe	host	2	***Sp. humboldti* (TY)**; *Botaurus sinensis* (CB)	[37]; this study
		vector	3	*Cx. pipiens* (NI); *Cx. inatomii* (NI)	[41]
pEUDCHR02	Japan only	host	1	***Es. chrysocome* (TY)**	MalAvi
pGALLUS02	Japan and Thailand	host	4	***Es. chrysocome* (NS)**; *Crossoptilon crossoptilon* (KN); *Streptopelia orientals* (KN)	[42]; MalAvi
		vector	2		
pGRW04 *	all except Antarctica	host	91	***Sp. humboldti* (TY, YA, KG)**; *Hypsipetes amaurotis* (TY); *Horornis diphone* (TY); *Zosterops japonicus* (TY); *Monticola solitarius* (TY)	[19,43]; MalAvi; this study
		vector	3	*Cx. pipiens* (TY, KN, NI); *Cx. quinquefasciatus* (OK)	[36,39,41,44]
pGRW06 *	all continents except Antarctica	host	103	***Sp. humboldti* (NS)**; ***Sp. magellanicus* (NI)**; *Hypsipetes amaurotis* (TY); *Horornis diphone* (TY); *Zosterops japonicus* (TY); *Troglodytes troglodytes* (ST); *Zoothera aurea* (TY); *Monticola solitarius* (TY); *Coccothraustes coccothraustes* (TY); *Chloris sinica* (TY)	[43,45]; this study
		vector	3		
pGRW11 *	Asia, Europe, Africa	host	55	***Sp. humboldti* (KN)**; ***Sp. magellanicus* (YA)**; ***Sp. demersus* (TY)**; ***Es. chrysolophus* (YA)**	[19]; MalAvi; this study
		vector	2	*Cx. pipiens* (TY); *Cx. quinquefasciatus* (OK)	[44,46]
pLINN1	Asia, Australia, Europe, North America	host	32	***Sp. humboldtii* (TY)**	MalAvi
		vector	6		
pLINOLI01	Asia, Europe, Africa	host	28	***Sp. demersus* (TY)**	MalAvi
pNYCNYC02 *	Japan only	host	5	***Sp. humboldti* (TY, KY, OS, OI)**; ***Sp. magellanicus* (IS, HG, OI)**; *Nycticorax nycticorax* (CB); *Luscinia cyanura* (ST); *Fringilla montifringilla*	[37,47]; MalAvi; this study
pPADOM02 *	Asia, Australia, Europe, Africa, North America	host	20	***El. minor* (TY)**; ***Sp. magellanicus* (OI)**; *Corvus corone*; *Passer montanus* (CB)	[37,48]; MalAvi; this study
		vector	6	*Ae. albopictus* (KN, NS); *Cx. bitaeniorhynchus* (NS); *Cx. inatomii* (NS); *Cx. pipiens* (TY, KN, NI); *Tr. bambusa* (KN); *Lt. vorax* (KN)	[17,29,38,39,46]
pSGS1 *	all continents except Antarctica	host	151	***Sp. humboldti* (TY, KN, KY, OS, YA, FO)**; ***Sp. magellanicus* (YN)**; ***Sp. demersus* (TY, KY)**; ***Es. chrysolophus* (TY, YA)**; *Hypsipetes amaurotis* (CB); *Spodiopsar cineraceus* (CB, KN)	[19,37]; MalAvi; this study
		vector	9	*Ae. albopictus* (KN); *Cx. pipiens* (TY, KN); *Cx. sasai* (TY); *Lt. vorax* (KN); *Cu. sigaensis* (KN)	[17,30,36,39,40,49]
pSPHUM01	Japan only	host	1	***Sp. humboldtii* (TY)**	MalAvi
pSPHUM02	Japan only	host	1	***Sp. humboldtii* (TY)**	MalAvi
pSPHUM03 *	Japan only	host	3	***Sp. humboldti* (NI, SZ)**; ***Sp. demersus* (SZ)**; *Calonectris leucomelas* (NI)	[29]; MalAvi; this study
pSPHUM05 *	Japan only	host	1	***Sp. humboldti* (HK, MG, NI)**	[29]; this study
		vector	1	*Cx. pipiens* (NI)	[29]

Table 2. *Cont.*

Lineages [a]	Distribution		Species [b]	Detected Species in Japan [c]	Reference
pSW5	Asia, Europe, Africa, North America	host	23	*Anas platyrhynchos* (HK, CB); *Fulica atra* (TY, CB); *Grus japonensis* (HK); *Podiceps cristatus* (CB); *Gallinago megala* (IB, CB, OK); *Gallinago hardwickii* (IB); ***Sp. magellanicus* (HG)**; *Calonectris leucomelas* (TY); *Botaurus eurhythmus* (NS)	[19,37,38,50,51]
pSYCON02 *	Japan and Spain	host	3	***Sp. humboldti* (KO)**; ***Sp. magellanicus* (OI)**	this study
		vector	1	*Cx. pipiens* (KN)	[39]
pTURPAL01 *	Japan only	host	2	***Py. papua* (HK)**; *Turdus pallidus* (OK)	MalAvi; this study
hAPPAT01	Japan only	host	1	***Ap. patagonicus* (HK)**	MalAvi
hHYPHI07	Asia	host	5	***Es. chrysocome* (HK)**; *Hypsipetes amaurotis* (CB); *Phylloscopus borealoides* (CB)	[37]; MalAvi
hPYGPAP01 *	Japan only	host	1	***Py. papua* (HK)**	this study
hSPHUM04	Japan only	host	1	***Sp. humboldtii* (KN)**	MalAvi
hSPHUM06 *	Japan only	host	1	***Sp. humboldtii* (SZ)**	this study
hSPMAG12 *	Japan only	host	5	*Larus crassirostris* (CB); ***Sp. humboldti* (TY, NI)**; ***Sp. magellanicus* (MG, IS, OI)**; ***Sp. demersus* (SZ, KY, OS)**; ***Es. chrysocome* (NI)**	[28,29]; MalAvi; this study

[a] Lineage names are according to the MalAvi database. The small case letter in front of the lineages represents the parasite genus (p = *Plasmodium*, h = *Haemoproteus*). Asterisks denote lineages detected in this study. [b] Number of species in which each lineage was detected, based on the MalAvi database (as of 14 January 2025) plus new host species found in this study. Lineages with no "vector" row have not been detected from any vector species. [c] Bold: penguin (Spheniscidae) species. Underline: detections from this study. Some genera are abbreviated: *Ap.* = *Aptenodytes*, *El.* = *Eudyptula*, *Sp.* = *Spheniscus*, *Es.* = *Eudyptes*, *Ae.* = *Aedes*, *Cx.* = *Culex*, *Tr.* = *Tripteroides*, *Lt.* = *Lutzia*. Parentheses denote the prefecture of detection: HK = Hokkaido, MG = Miyagi, IB = Ibaraki, ST = Saitama, CB = Chiba, TY = Tokyo, KN = Kanagawa, NI = Niigata, KY = Kyoto, HG = Hyogo, YA = Yamaguchi, FO = Fukuoka, NS = Nagasaki, OI = Oita, KG = Kagoshima, OK = Okinawa. [d] Unpublished records directly mined from MalAvi.

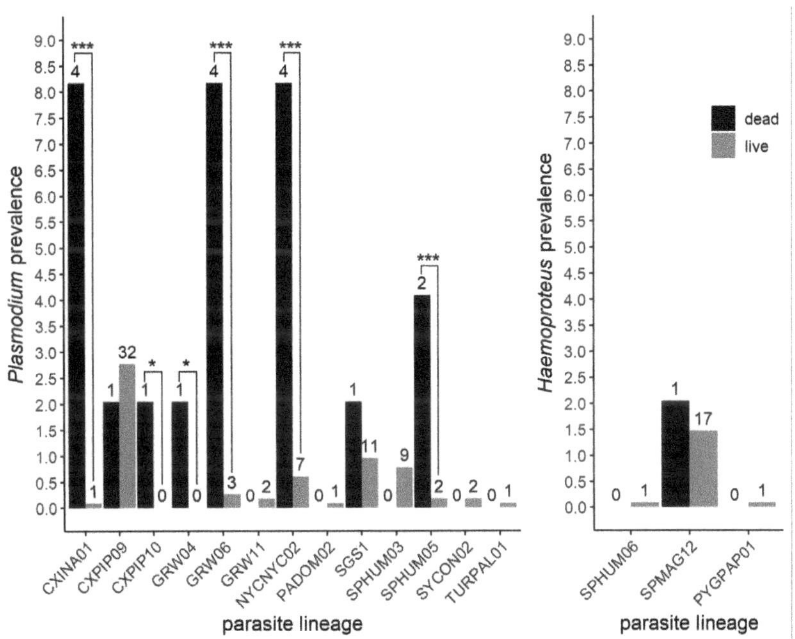

Figure 2. *Plasmodium* and *Haeamoproteus* prevalence by parasite lineage. Dark bars = dead individuals. Light bars = live individuals. Numbers above the bars are the number of positive individuals. Asterisks denote significant differences in prevalence between dead and live penguins (*: $0.01 \leq p < 0.05$, ***: $p < 0.01$).

In 6 of the 13 *Plasmodium* lineages, the odds of death were significantly higher among infected birds compared to non-infected birds (Fisher's exact test: CXINA01 OR = 94.25 $p < 0.01$, CXPIP10 $p = 0.04$, GRW04 $p = 0.04$, GRW06 OR = 31.53 $p < 0.01$, NYCNYC02 OR = 13.53 $p < 0.01$, SPHUM05 OR = 22.77 $p = 0.01$). In all other lineages, significant differences were not detected (SGS1 OR = 2.03 $p = 0.41$, SPMAG12 OR = 1.31 $p = 0.55$, all other lineages $p = 1$). Note that OR could not be calculated in some lineages because of zero values in one group.

3.3. Haemosporidian Prevalence by Microscopy and Parasitemia

By microscopy, the overall prevalence was 3.88% ($n = 37$) for *Plasmodium* parasites and 0.73% ($n = 7$) for *Haemoproteus* parasites (Table 3 and Table S2). Among individuals for which blood smears were prepared, the PCR prevalence was 5.67% (*Plasmodium*, $n = 54$) and 1.47% (*Haemoproteus*, $n = 14$), showing no significant difference in comparison to the prevalence by microscopy (Fisher's exact test: *Plasmodium* $p = 0.09$, *Haemoproteus* $p = 0.19$). One Magellanic penguin was positive for *Plasmodium* parasites by microscopy but negative by PCR. All other individuals positive by microscopy were also positive by PCR.

For individuals in which *Plasmodium* parasites were detected microscopically, the parasitemia was significantly higher in deceased individuals compared to live individuals (*t*-test: $p = 0.02$ t = −3.62 df = 4.02, GRW06 only $p = 0.05$, t = −3.07, df = 3.01; Figure 3).

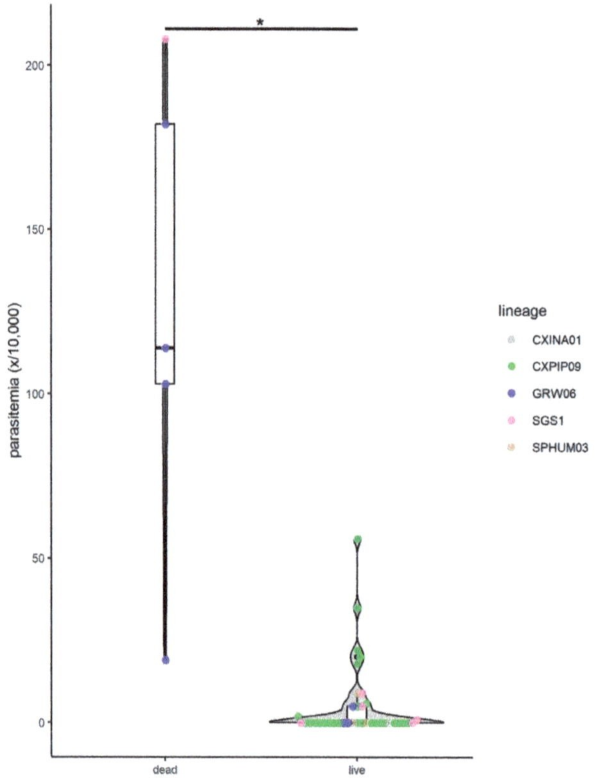

Figure 3. Parasitemia in *Plasmodium*-infected penguins, by vital status. Colors show different lineages, as shown in the legend. If an infected erythrocyte was found outside of the counted 10,000 erythrocytes, the parasitemia was calculated as 1/100,000 erythrocytes. The asterisk denotes a significant difference in parasitemia ($p < 0.05$).

Table 3. Summary of haemosporidian detection in penguins of this study in which blood smears were available.

Parasite Genus	Species [a]	Live				Deceased (All Outdoor)		Total	
		Indoor		Outdoor					
		Individuals	Positive [b]	Individuals	Positive [b]	Individuals	Positive [b]	Individuals	Positive [b]
Plasmodium	King penguin (*Ap. patagonicus*)	33	0/0	1	0/0			34	0/0
	Emperor penguin (*Ap. forsteri*)	2	0/0					2	0/0
	Adelie penguin (*Py. adeliae*)	14	0/0					14	0/0
	Chinstrap penguin (*Py. antarcticus*)	15	0/0					15	0/0
	Gentoo penguin (*Py. papua*)	61	0/0	1	0/0			62	0/0
	Fairy penguin (*El. minor*)			2	0/0			2	0/0
	Humboldt penguin (*Sp. humboldti*)	21	0/0	446	21/9	5	5/5	472	26/14
	Magellanic penguin (*Sp. magellanicus*)	58	1/1	84	14/10	1	0/0	143	15/11
	African penguin (*Sp. demersus*)	19	0/0	147	13/12	1	0/0	167	13/12
	N. rockhopper penguin (*Es. moseleyi*)	23	0/0	7	0/0			30	0/0
	S. rockhopper penguin (*Es. chrysocome*)	8	0/0	4	0/0			12	0/0
	total	254	1/1	692	48/31	7	5/5	953	54/37
Haemoproteus	King penguin (*Ap. patagonicus*)	33	0/0	1	0/0			34	0/0
	Emperor penguin (*Ap. forsteri*)	2	0/0					2	0/0
	Adelie penguin (*Py. adeliae*)	14	0/0					14	0/0
	Chinstrap penguin (*Py. antarcticus*)	15	0/0					15	0/0
	Gentoo penguin (*Py. papua*)	61	0/0	1	0/0			62	0/0
	Fairy penguin (*El. minor*)			2	0/0			2	0/0
	Humboldt penguin (*Sp. humboldti*)	21	0/0	446	1/0	5	0/0	472	1/0
	Magellanic penguin (*Sp. magellanicus*)	58	0/0	84	7/4	1	0/0	143	7/4
	African penguin (*Sp. demersus*)	19	1/0	147	4/2	1	0/0	167	5/2
	N. rockhopper penguin (*Es. moseleyi*)	23	0/0	7	0/0			30	0/0
	S. rockhopper penguin (*Es. chrysocome*)	8	0/0	4	1/1			12	1/1
		254	1/0	692	13/7	7	0/0	953	14/7

[a] Genera are abbreviated: *Ap.* = *Aptenodytes*, *Py.* = *Pygoscelis*, *El.* = *Eudyptula*, *Sp.* = *Spheniscus*, *Es.* = *Eudyptes*. [b] Left = PCR results; Right = microscopy results.

4. Discussion

While there have been sporadic reports of avian haemosporidia in Japan, only few facilities have been surveyed [19,27,29,52]. This is the first collective study to have surveyed avian haemosporidia in captive penguins throughout Japan. Parasites were detected from penguins in all eight areas of Japan, displaying the widespread distribution of avian haemosporidia amongst captive penguins. *Leucocytozoon* parasites were not detected in this study, possibly due to the relatively high specificity known in this genus or the lack of environments favored by vector blackflies [1,12,53]. Although information on relocation history was not available for many individuals, all except the two novel lineages have previously been detected from wild birds and/or vector insects of Japan, strongly suggesting that all individuals were infected in Japan.

The *Plasmodium* prevalence was significantly higher for the total prevalence compared to initial prevalence. Although not significant, eight initially negative individuals were found to be *Haemoproteus*-positive later on. As circannual fluctuations in parasitemia and medications may also be involved, it is difficult to confirm whether new infections were acquired [29]. Mosquitoes have been confirmed at zoological facilities [17,29,30]. While few investigations on biting midges have been carried out at zoological facilities in Japan, *Culicoides* biting midges have been confirmed at a facility in the Chubu area (pers. comm.). This suggests that new infections are possible at such zoological facilities, and along with the possibility of circannual fluctuations, it is important to perform periodical check-ups to investigate haemosporidian infection status.

As widely suggested, outdoor enclosures critically increase the exposure of vector insects to penguins and therefore the risk of avian malaria [18,54], as shown by data from this study. These results re-confirm that the ultimate and most effective prevention is to keep all penguins in indoor facilities. However, it is important to understand that not all facilities are capable of such environments, and other preventative measures such as medication and air curtains cannot be neglected.

Plasmodium infection can be highly lethal for penguins [18,54], as the odds of death were significantly higher for *Plasmodium*-infected individuals. Furthermore, the odds of death differed between parasite lineages, similar to previous studies in the US and UK [55,56]. Of the four lineages with particularly high odds of death (pCXINA01, pGRW06, pNYCNYC02, and pSPHUM05), all except GRW06 have only been detected in Japan and little information is known. GRW06 of *P. elongatum*, which is one of the most widespread lineages of *Plasmodium* spp., has been associated with mortality in penguins in New Zealand and Brazil [13,57,58]. This study adds evidence on the virulence of GRW06 in penguins. Furthermore, the *P. elongatum* parasitemia was significantly higher in deceased penguins, which was also seen in a previous study [57]. Although lethal cases with low parasitemia have also been reported [18], GRW06 might require high parasitemia in order to exhibit lethality. However, note that findings in this study may be a result of many factors such as sampling bias, medication, co-infections by other pathogens, and physiological factors [13,55,56].

Previous studies have suggested weak connections between *Haemoproteus* parasites and penguin mortality [12,59,60]. A previous study reports a lethal case in penguins due to haemoproteosis [27] but is considered problematic [28]. Meanwhile, gametocytes of SPMAG12 were detected from individuals including those that show no symptomatic signs, potentially related to the extremely low parasitemia [28]. The slightly high odds of death seen in this study may therefore be due to other factors rather than solely *Haemoproteus* infection.

There were several limitations upon data analysis of this study. Comparisons between penguin species were not possible, due to the strong bias between the number of individuals and environments. Like in the UK [54], some of the most abundant species had the highest parasite prevalence. Furthermore, some species such as chinstrap penguins (*Pygoscelis antarcticus*) were kept exclusively indoors, while others such as Humboldt penguins tended to be kept outdoors, in relation to their natural environment. As found in this study, the captivity environment (indoors or outdoors) was a significant factor for haemosporidian prevalence, and the unbalanced numbers of each penguin species in each environment made analysis by species impossible. Additionally, information such as birth location, relocation history, medication history, and symptoms were not available for many individuals. Such information would make further analyses such as location of infection and virulence possible. Nonetheless, comprehensive studies such as this study are important to better understand the status of infection in penguins, which are crucial for preventative measures and ultimately the conservation of penguin populations. Furthermore, only roughly half of the facilities in Japan which keep penguins were investigated in this study. Future studies at uninvestigated facilities as well as continuous monitoring at already-investigated facilities are anticipated, both within Japan and throughout the world.

Supplementary Materials: The following supporting information can be downloaded at: https://www.mdpi.com/article/10.3390/jzbg6010007/s1, Table S1: Summary of haemosporidian detection in penguins of this study, by prefecture, environmental factor, and vital status; Table S2: Summary of haemosporidian lineages detected from penguins in which blood smears were available; Figure S1: Map of prefectures investigated in this study.

Author Contributions: Conceptualization, K.M. and Y.S.; methodology, K.M. and Y.S.; validation, M.I., K.M. and Y.S.; formal analysis, M.I.; investigation, M.I., M.S. and A.S.; resources, K.M. and Y.S.; data curation, M.I., M.S. and A.S.; writing—original draft preparation, M.I., M.S. and A.S.; writing—review and editing, K.M. and Y.S.; visualization, M.I., M.S. and A.S.; supervision, K.M. and Y.S.; project administration, K.M. and Y.S.; funding acquisition, Y.S. All authors have read and agreed to the published version of the manuscript.

Funding: This research was partially funded by the Strategic Research Base Development Program "International Research on the Management of Zoonosis in Globalization and Training of Young Researchers" from the MEXT of Japan, project no. S1491007, and the Grant-in-Aid for Scientific Research from the Japan Society for the Promotion of Science (JSPS), KAKENHI nos. 21580406, 26450484, 19J20367, 21K05961 and 24K09222.

Institutional Review Board Statement: Ethical review and approval were waived for this study as the samples obtained were collected and provided by veterinarians who routinely collect blood samples for each health check of the captive penguins.

Data Availability Statement: The original contributions presented in this study are included in the article/Supplementary Material. Additionally, data on the detected parasite lineages have been deposited in the MalAvi database (http://130.235.244.92/Malavi/ (accessed on 14 January 2025)).

Acknowledgments: We are grateful to the staff of all facilities that provided samples for this study. We also thank former members of the Laboratory of Biomedical Science at Nihon University—S. Aratani, M. Horinouchi, N. Oda, A. Ono, Y. Osawa, A. Takasu, and M. Yamamoto—who helped in analyzing the penguin samples.

Conflicts of Interest: The authors declare no conflicts of interest.

References

1. Valkiūnas, G. *Avian Malaria Parasites and Other Haemosporidia*; CRC Press: Boca Raton, FL, USA, 2005; Volume 1, ISBN 9788578110796.
2. Atkinson, C.T.; Dusek, R.J.; Woods, K.L.; Iko, W.M. Pathogenicity of Avian Malaria in Experimentally-Infected Hawaii Amakihi. *J. Wildl. Dis.* **2000**, *36*, 197–204. [CrossRef] [PubMed]
3. Møller, A.P.; Erritzøe, J. Host Immune Defence and Migration in Birds. *Evol. Ecol.* **1998**, *12*, 945–953. [CrossRef]

4. Warner, R.E. The Role of Introduced Diseases in the Extinction of the Endemic Hawaiian Avifauna. *Condor* **1968**, *70*, 101–120. [CrossRef]
5. Van Riper, C.I.; Van Riper, S.G.; Goff, M.L.; Laird, M. The Epizootiology and Ecological Significance of Malaria in Hawaiian Land Birds. *Ecol. Monogr.* **1986**, *56*, 327–344. [CrossRef]
6. Jia, T.; Huang, X.; Valkiūnas, G.; Yang, M.; Zheng, C.; Pu, T.; Zhang, Y.; Dong, L.; Suo, X.; Zhang, C. Malaria Parasites and Related Haemosporidians Cause Mortality in Cranes: A Study on the Parasites Diversity, Prevalence and Distribution in Beijing Zoo. *Malar. J.* **2018**, *17*, 234. [CrossRef] [PubMed]
7. Ortiz-Catedral, L.; Brunton, D.; Stidworthy, M.F.; Elsheikha, H.M.; Pennycott, T.; Schulze, C.; Braun, M.J.; Wink, M.; Gerlach, H.; Pendl, H.; et al. *Haemoproteus Minutus* Is Highly Virulent for Australasian and South American Parrots. *Parasit. Vectors* **2019**, *12*, 40. [CrossRef] [PubMed]
8. Olias, P.; Wegelin, M.; Zenker, W.; Freter, S.; Gruber, A.D.; Klopfleisch, R.; Klopflisch, R. Avian Malaria Deaths in Parrots, Europe. *Emerg. Infect. Dis.* **2011**, *17*, 950–952. [CrossRef] [PubMed]
9. Donovan, T.A.; Schrenzel, M.; Tucker, T.A.; Pessier, A.P.; Stalis, I.H. Hepatic Hemorrhage, Hemocoelom, and Sudden Death Due to *Haemoproteus* Infection in Passerine Birds: Eleven Cases. *J. Vet. Diagn. Investig.* **2008**, *20*, 304–313. [CrossRef]
10. Lee, S.-H.; Kwak, D.; Kim, K.-T. The First Clinical Cases of *Haemoproteus* Infection in a Snowy Owl (*Bubo scandiacus*) and a Goshawk (*Accipiter gentilis*) at a Zoo in the Republic of Korea. *J. Vet. Med. Sci.* **2018**, *80*, 1255–1258. [CrossRef] [PubMed]
11. Yoshimoto, M.; Ozawa, K.; Kondo, H.; Echigoya, Y.; Shibuya, H.; Sato, Y.; Sehgal, R.N.M. A Fatal Case of a Captive Snowy Owl (*Bubo scandiacus*) with *Haemoproteus* Infection in Japan. *Parasitol. Res.* **2020**, *120*, 277–288. [CrossRef] [PubMed]
12. Vanstreels, R.E.T.; Braga, É.M.; Catão-Dias, J.L. Blood Parasites of Penguins: A Critical Review. *Parasitology* **2016**, *143*, 931–956. [CrossRef]
13. Ings, K.; Denk, D. Avian Malaria in Penguins: Diagnostics and Future Direction in the Context of Climate Change. *Animals* **2022**, *12*, 600. [CrossRef] [PubMed]
14. IUCN. The IUCN Red List of Threatened Species Version 2020-1. Available online: www.iucnredlist.org (accessed on 1 January 2025).
15. Schneider, T.; Olsen, D.; Dykstra, C.; Huettner, S.; Branch, S.; Sirpenski, G.; Sarro, S.; Waterfall, K.; Henry, L.; DuBois, L.; et al. *Penguin (Spheniscidae) Care Manual*; Association of Zoos and Aquariums: Silver Spring, MD, USA, 2014.
16. Bueno, M.G.; Lopez, R.P.G.; de Menezes, R.M.T.; de Jesus Costa-Nascimento, M.; de Castro Lima, G.F.M.; de Sousa Araújo, R.A.; Vaz Guida, F.J.; Kirchgatter, K. Identification of *Plasmodium relictum* Causing Mortality in Penguins (*Spheniscus magellanicus*) from São Paulo Zoo, Brazil. *Vet. Parasitol.* **2010**, *173*, 123–127. [CrossRef]
17. Ejiri, H.; Sato, Y.; Sawai, R.; Sasaki, E.; Matsumoto, R.; Ueda, M.; Higa, Y.; Tsuda, Y.; Omori, S.; Murata, K.; et al. Prevalence of Avian Malaria Parasite in Mosquitoes Collected at a Zoological Garden in Japan. *Parasitol. Res.* **2009**, *105*, 629–633. [CrossRef] [PubMed]
18. Grilo, M.; Vanstreels, R.E.T.; Wallace, R.; García-Párraga, D.; Braga, É.M.; Chitty, J.; Catão-Dias, J.L.; Madeira de Carvalho, L.M. Malaria in Penguins – Current Perceptions. *Avian Pathol.* **2016**, *45*, 393–407. [CrossRef] [PubMed]
19. Tamura, J.; Sumiyama, D.; Kanazawa, T.; Sato, Y.; Murata, K. A Study on Molecular Phylogeny of the Avian Malaria Parasite's MSP1 Gene from the Penguins in Captivity (Japanese). *J. Jpn. Assoc. Zoos Aquar.* **2018**, *60*, 103–111.
20. Shimazu, M.; Tatara, S.; Kinuta, T. Avian Malaria (*Plasmodium* sp.) in Magellanic Penguins (*Spheniscus magellanicus*) and a Rockhopper Penguin (*Eudyptes chrysocome*) in Captivity (Japanese). *J. Jpn. Assoc. Zoos Aquar.* **1994**, *35*, 107–117.
21. Fix, A.S.; Waterhouse, C.; Greiner, E.C.; Stoskopf, M.K. *Plasmodium relictum* as a Cause of Avian Malaria in Wild-Caught Magellanic Penguins (*Spheniscus magellanicus*). *J. Wildl. Dis.* **1988**, *24*, 610–619. [CrossRef] [PubMed]
22. Jones, H.I.; Shellam, G.R. Blood Parasites in Penguins, and Their Potential Impact on Conservation. *Mar. Ornithol.* **1999**, *27*, 181–184.
23. Scott, H.H. Report on the Deaths Occurring in the Society's Gardens during the Year 1926. *Proc. Zool. Soc. London* **1927**, *1*, 173–198. [CrossRef]
24. Grilo, M. Characterization of Infection by Malaria Parasites in Penguins Housed in Zoological Collections. Maste's Thesis, Universidade de Lisboa, Lisboa, Portugal, 2014.
25. Murata, M.; Murakami, M. Two Distinct MtDNA Lineages among Captive African Penguins in Japan. *J. Vet. Med. Sci.* **2014**, *76*, 559–563. [CrossRef] [PubMed]
26. Takami, K. Consideration on the Vaccination to Captive Rare and Endangered Species Based on Their Conditions (Japanese). *Jpn. J. Zoo Wildl. Med.* **2012**, *17*, 63–71. [CrossRef]
27. Ishikawa, H.; Hasegawa, K. Protozoiasis Suspected as Haemoproteus Sp. in Magellanic Penguins Spheniscus Magellanics in Captivity (Japanese). *Toba Aquar. Annu. Rep.* **1989**, *1*, 77–89.
28. Inumaru, M.; Aratani, S.; Shimizu, M.; Yamamoto, M.; Sato, Y.; Murata, K.; Valkiūnas, G. Penguins Are Competent Hosts of *Haemoproteus* Parasites: The First Detection of Gametocytes, with Molecular Characterization of *Haemoproteus Larae*. *Parasit. Vectors* **2020**, *13*, 307. [CrossRef]

29. Inumaru, M.; Yamada, A.; Shimizu, M.; Ono, A.; Horinouchi, M.; Shimamoto, T.; Tsuda, Y.; Murata, K.; Sato, Y. Vector Incrimination and Transmission of Avian Malaria at an Aquarium in Japan: Mismatch in Parasite Composition between Mosquitoes and Penguins. *Malar. J.* **2021**, *20*, 136. [CrossRef]
30. Ejiri, H.; Sato, Y.; Kim, K.S.; Hara, T.; Tsuda, Y.; Imura, T.; Murata, K.; Yukawa, M. Entomological Study on Transmission of Avian Malaria Parasites in a Zoological Garden in Japan: Bloodmeal Identification and Detection of Avian Malaria Parasite DNA from Blood-Fed Mosquitoes. *J. Med. Entomol.* **2011**, *48*, 600–607. [CrossRef] [PubMed]
31. Hellgren, O.; Waldenström, J.; Bensch, S. A New PCR Assay for Simultaneous Studies of *Leucocytozoon*, *Plasmodium*, and *Haemoproteus* from Avian Blood. *J. Parasitol.* **2004**, *90*, 797–802. [CrossRef] [PubMed]
32. NCBI Resource Coordinators Database Resources of the National Center for Biotechnology Information. *Nucleic Acid Res.* **2018**, *46*, D8–D13. [CrossRef]
33. Madden, T. The BLAST Sequence Analysis Tool. In *The NCBI Handbook*; National Center for Biotechnology Information: Bethesda, MD, USA, 2013.
34. Bensch, S.; Hellgren, O.; Pérez-Tris, J. MalAvi: A Public Database of Malaria Parasites and Related Haemosporidians in Avian Hosts Based on Mitochondrial Cytochrome *b* Lineages. *Mol. Ecol. Resour.* **2009**, *9*, 1353–1358. [CrossRef] [PubMed]
35. R Core Team. R: A Language and Environment for Statistical Computing. 2020. Available online: https://www.r-project.org/ (accessed on 10 January 2024).
36. Kim, K.S.; Tsuda, Y. Sporogony and Sporozoite Rates of Avian Malaria Parasites in Wild *Culex Pipiens Pallens* and *C. Inatomii* in Japan. *Parasit. Vectors* **2015**, *8*, 633. [CrossRef] [PubMed]
37. Inumaru, M.; Murata, K.; Sato, Y. Prevalence of Avian Haemosporidia among Injured Wild Birds in Tokyo and Environs, Japan. *Int. J. Parasitol. Parasites Wildl.* **2017**, *6*, 299–309. [CrossRef]
38. Tanigawa, M.; Sato, Y.; Ejiri, H.; Imura, T.; Chiba, R.; Yamamoto, H.; Kawaguchi, M.; Tsuda, Y.; Murata, K.; Yukawa, M. Molecular Identification of Avian Haemosporidia in Wild Birds and Mosquitoes on Tsushima Island, Japan. *J. Vet. Med. Sci.* **2013**, *75*, 319–326. [CrossRef] [PubMed]
39. Odagawa, T.; Inumaru, M.; Sato, Y.; Murata, K.; Higa, Y.; Tsuda, Y. A Long-Term Field Study on Mosquito Vectors of Avian Malaria Parasites in Japan. *J. Vet. Med. Sci.* **2022**, *84*, 1391–1398. [CrossRef] [PubMed]
40. Kim, K.S.; Tsuda, Y.; Sasaki, T.; Kobayashi, M.; Hirota, Y. Mosquito Blood-Meal Analysis for Avian Malaria Study in Wild Bird Communities: Laboratory Verification and Application to *Culex Sasai* (Diptera: Culicidae) Collected in Tokyo, Japan. *Parasitol. Res.* **2009**, *105*, 1351–1357. [CrossRef]
41. Kim, K.S.; Tsuda, Y. Avian *Plasmodium* Lineages Found in Spot Surveys of Mosquitoes from 2007 to 2010 at Sakata Wetland, Japan: Do Dominant Lineages Persist for Multiple Years? *Mol. Ecol.* **2012**, *21*, 5374–5385. [CrossRef] [PubMed]
42. Murata, K.; Nii, R.; Sasaki, E.; Ishikawa, S.; Sato, Y.; Sawabe, K.; Tsuda, Y.; Matsumoto, R.; Suda, A.; Ueda, M. *Plasmodium (Bennettinia) Juxtanucleare* Infection in a Captive White Eared-Pheasant (*Crossoptilon crossoptilon*) at a Japanese Zoo. *J. Vet. Med. Sci.* **2008**, *70*, 203–205. [CrossRef] [PubMed]
43. Inumaru, M.; Kimura, R.; Suzuki, N.; Suzuki, H.; Horikoshi, K.; Nishiumi, I.; Kawakami, K.; Tsuda, Y.; Murata, K.; Sato, Y. Prevalence and Transmission Cycle of Avian Pathogens in the Isolated Oceanic Islands of Japan. *Ecol. Evol.* **2024**, *14*, e70737. [CrossRef] [PubMed]
44. Ejiri, H.; Sato, Y.; Sasaki, E.; Sumiyama, D.; Tsuda, Y.; Sawabe, K.; Matsui, S.; Horie, S.; Akatani, K.; Takagi, M.; et al. Detection of Avian *Plasmodium* Spp. DNA Sequences from Mosquitoes Captured in Minami Daito Island of Japan. *J. Vet. Med. Sci.* **2008**, *70*, 1205–1210. [CrossRef]
45. Imura, T.; Suzuki, Y.; Ejiri, H.; Sato, Y.; Ishida, K.; Sumiyama, D.; Murata, K.; Yukawa, M. Prevalence of Avian Haematozoa in Wild Birds in a High-Altitude Forest in Japan. *Vet. Parasitol.* **2012**, *183*, 244–248. [CrossRef]
46. Kim, K.S.; Tsuda, Y. Seasonal Changes in the Feeding Pattern of *Culex Pipiens Pallens* Govern the Transmission Dynamics of Multiple Lineages of Avian Malaria Parasites in Japanese Wild Bird Community. *Mol. Ecol.* **2010**, *19*, 5545–5554. [CrossRef] [PubMed]
47. Inumaru, M.; Nishiumi, I.; Kawakami, K.; Sato, Y. A Widespread Survey of Avian Haemosporidia in Deceased Wild Birds of Japan: The Hidden Value of Personally Collected Samples. *J. Vet. Med. Sci.* **2022**, *84*, 1253–1260. [CrossRef] [PubMed]
48. Beadell, J.S.; Ishtiaq, F.; Covas, R.; Melo, M.; Warren, B.H.; Atkinson, C.T.; Bensch, S.; Graves, G.R.; Jhala, Y.V.; Peirce, M.A.; et al. Global Phylogeographic Limits of Hawaii's Avian Malaria. *Proc. R. Soc. B Biol. Sci.* **2006**, *273*, 2935–2944. [CrossRef] [PubMed]
49. Inumaru, M.; Nakamura, K.; Odagawa, T.; Suzuki, M.; Murata, K.; Sato, Y. The First Detection of Avian Haemosporidia from *Culicoides* Biting Midges in Japan, with Notes on Potential Vector Species and the Transmission Cycle. *Vet. Parasitol. Reg. Stud. Reports* **2023**, *39*, 100840. [CrossRef] [PubMed]
50. Inumaru, M.; Odaya, Y.; Sato, Y.; Marzal, A. First Records of Prevalence and Diversity of Avian Haemosporidia in Snipe Species (Genus *Gallinago*) of Japan. *Int. J. Parasitol. Parasites Wildl.* **2021**, *16*, 5–17. [CrossRef] [PubMed]

51. Yoshimura, A.; Koketsu, M.; Bando, H.; Saiki, E.; Suzuki, M.; Watanabe, Y.; Kanuka, H.; Fukumoto, S. Phylogeny Comparison of Avian Haemosporidian Parasites from Resident and Migratory Birds in Northern Japan. *J. Wildl. Dis.* **2014**, *50*, 235–242. [CrossRef] [PubMed]
52. Yamada, M.; Takizawa, R. Incidence of Avian Malaria at an Animal Facility within the Prefecture (Translated from Japanese). *Proc. Oita Prefect. Livest. Hygene Livest. Relat. Perform. Present.* **2010**, *59*, 70–72.
53. Tongjura, J.; Amuga, G.; Mafuyai, H.; Matur, B.; Olatunwa, J. Influence of Some Water Physicochemical Parameters on the Distribution of Black Fly (Diptera, Simuliidae) in Some Rivers in Nasarawa State, Nigeria. *Adv. Entomol.* **2015**, *3*, 101–110. [CrossRef]
54. Hernandez-Colina, A.; Gonzalez-Olvera, M.; Eckley, L.; Lopez, J.; Baylis, M. Avian Malaria Affecting Penguins in Zoological Gardens, Aquariums and Wildlife Parks in the UK. *Vet. Rec.* **2021**, *189*, e511. [CrossRef] [PubMed]
55. Spottiswoode, N.; Bartlett, S.L.; Conley, K.J.; Seimon, T.A.; Griffin, D.O.; Sykes, J.M. Analysis of *Plasmodium* Lineages Identified in Captive Penguins (*Sphenisciformes* spp.), Eiders (*Somateria* spp.), and Inca Terns (*Larosterna inca*) in the North American Zoological Collection. *J. Zoo Wildl. Med.* **2020**, *51*, 140–149. [CrossRef] [PubMed]
56. Gonzalez-Olvera, M.; Hernandez-Colina, A.; Himmel, T.; Eckley, L.; Lopez, J.; Chantrey, J.; Baylis, M.; Jackson, A.P. Molecular and Epidemiological Surveillance of Plasmodium Spp. during a Mortality Event Affecting Humboldt Penguins (Spheniscus Humboldti at a Zoo in the UK. *Int. J. Parasitol. Parasites Wildl.* **2022**, *19*, 26–37. [CrossRef] [PubMed]
57. Vanstreels, R.E.T.; de Angeli Dutra, D.; Ferreira, F.C.J.; Hurtado, R.; Egert, L.; Mayorga, L.F.S.P.; Bhering, R.C.C.; Braga, É.M.; Catão-Dias, J.L. Epidemiology, Hematology, and Unusual Morphological Characteristics of Plasmodium during an Avian Malaria Outbreak in Penguins in Brazil. *Parasitol. Res.* **2019**, *118*, 3497–3508. [CrossRef] [PubMed]
58. Sijbranda, D.C.; Hunter, S.; Howe, L.; Lenting, B.; Argilla, L.; Gartrell, B.D. Cases of Mortality in Little Penguins (*Eudyptula minor*) in New Zealand Associated with Avian Malaria. *N. Z. Vet. J.* **2017**, *65*, 332–337. [CrossRef] [PubMed]
59. Sallaberry-Pincheira, N.; Gonzalez-Acuña, D.; Herrera-Tello, Y.; Dantas, G.P.M.; Luna-Jorquera, G.; Frere, E.; Valdés-Velasquez, A.; Simeone, A.; Vianna, J.A. Molecular Epidemiology of Avian Malaria in Wild Breeding Colonies of Humboldt and Magellanic Penguins in South America. *Ecohealth* **2015**, *12*, 267–277. [CrossRef]
60. Levin, I.I.; Outlaw, D.C.; Vargas, F.H.; Parker, P.G. *Plasmodium* Blood Parasite Found in Endangered Galapagos Penguins (*Spheniscus mendiculus*). *Biol. Conserv.* **2009**, *142*, 3191–3195. [CrossRef]

Journal of *Zoological and Botanical Gardens*

MDPI

Article

Gastrointestinal Parasites in Reptiles from a Portuguese Zoo

David Dantas [1], Cláudia Luís Batista [1], Maria João Castro [1], Nuno Alvura [2] and Teresa Letra Mateus [3,4,5,*]

1 Escola Superior Agrária, Instituto Politécnico de Viana do Castelo, 4990-706 Ponte de Lima, Portugal; davidantas1998@gmail.com (D.D.); enfvetclaudia@gmail.com (C.L.B.); mariajoaocastro@sapo.pt (M.J.C.)
2 Zoo da Maia, Rua da Estação, 4470-184 Maia, Portugal
3 CISAS—Center for Research and Development in Agrifood Systems and Sustainability, Escola Superior Agrária, Instituto Politécnico de Viana do Castelo, Rua Escola Industrial e Comercial de Nun'Àlvares, 4900-347 Viana do Castelo, Portugal
4 EpiUnit ITR, Instituto de Saúde Pública, Universidade do Porto, Rua das Taipas, nº 135, 4050-091 Porto, Portugal
5 Associate Laboratory for Animal and Veterinary Sciences (AL4AnimalS) Quinta de Prados, Veterinary and Animal Research Centre (CECAV), Universidade de Trás-os-Montes e Alto Douro, 5000-801 Vila Real, Portugal
* Correspondence: tlmateus@esa.ipvc.pt

Abstract: The growing popularity of reptiles has contributed to their reproduction in captivity. When subjected to stressful environments, such as the presence of a higher number of humans and animals, reptiles may become more susceptible to parasites. Endoparasites in captive animals may cause several clinical signs ranging from mild to severe: lethargy, anorexia, diarrhea, cloacal/penile prolapse, infertility, intestinal malabsorption syndrome, and weight loss, among others. This study aimed to assess the presence of gastrointestinal parasites in fecal samples of reptiles from a Portuguese zoo through two techniques: a fecal flotation test (using a saturated sodium chloride solution) and Mini-FLOTAC. Ninety-nine samples belonging to 22 different animal species were collected and analyzed. Parasites were identified in 53.5% of the samples. Chelonians had a higher frequency (100%), followed by lizards (56.8%) and snakes (47.4%). The eggs/oocysts found were oxyurids (36.4%), strongylids/*Kalicephalus* sp. (8.1%), *Eimeria* sp. (5.1%), *Hymenolepis* spp. (5.1%), ascarids (4.0%), and *Isospora* sp. (2.0%). Both techniques presented the same results for each sample. The high prevalence of oxyurids, as well as of other parasites, can be explained by possible environmental contamination as these reptiles are kept in captivity. This study indicated the importance of assessing parasitic infections in reptiles in zoos, where routine coprological examinations should always be considered, as well as adequate prophylaxis.

Keywords: endoparasites; *Hymenolepis*; *Lampropeltis getula californiae*; oxyurid; One Health; *Pogona vitticeps*; strongylid

1. Introduction

The growing popularity of reptiles has contributed to their captive breeding. However, there are few studies regarding the prevalence of diseases, particularly parasitic ones, which affect these animals in zoos. Reptiles, when subjected to stressful environments (such as those with a high concentration of animals and humans in restricted habitats), can become more immunologically susceptible to parasites that are often pathogenic to them, including possibly to other animals, such as humans [1,2]. Therefore, this topic has to be investigated using a One Health perspective. Reptiles kept in captivity have shown a higher prevalence of parasites and more efficient transmission of various monoxenous parasites compared to wild reptiles [2,3]. Additionally, in captive animals that were captured from their natural habitats, the absence of intermediate hosts, dietary changes, and other factors have been demonstrated to promote the adaptation of parasites to new conditions, potentially threatening other zoo animals, since there are different susceptibilities among different reptiles [1].

Endoparasites can cause different clinical outcomes depending on the species and their life cycle, habitat conditions, infection severity, and the host (age, sex, and health) [3]. Although mild helminth infections are generally well tolerated by reptiles, severe infections can result in clinical pathology, especially in young or immunocompromised animals [4]. An oxyurid infection may lead to lethargy, anorexia, diarrhea, cloacal/penile prolapse, growth disturbances, infertility, and, in high burdens, it can cause intestinal malabsorption syndrome, which can result in severe clinical conditions [5,6]. *Kalicephalus* sp. is the most common nematode in snakes and may cause weight loss, enteritis, lethargy, and anorexia [7]. Like helminths, reptiles also host a variety of intestinal unicellular pathogens, such as *Blastocystis* spp., *Cryptosporidium* spp., *Eimeria* spp., *Entamoeba* spp., *Giardia* spp., and *Isospora* spp. [8]. An infection by *Entamoeba invadens* can cause anorexia, dehydration, bloody diarrhea, hepatitis, gastritis, and colitis. Conversely, an infection by *Eimeria* spp. can result in weight loss and enteritis [4,9]. In Table 1, some epidemiological studies from European countries (with special emphasis on Portugal), with the prevalence of different helminths and protozoa identified in fecal samples from reptiles, are presented.

Table 1. Prevalence (%) of helminths and protozoa identified in fecal samples from reptiles.

Countries	Reptiles	Protozoa	%	Cestodes	%	Nematodes	%	Reference
Several countries	Snakes	*Cryptosporidium* sp. *Cyclospora* sp. *Nyctotherus* sp. *Tetratrichomonas* sp.	1.9 1.9 1.9 1.8	Unidentified	-	Strongylida Ascarididae *Strongyloides* sp. *Capillaria* sp. Oxyuridae	20.4 7.4 5.6 3.7 1.9	[2]
	Lizards	*Nyctotherus* sp. *Balantidium* sp. *Cryptosporidium* sp. *Isospora* sp. *Eimeria* sp. *Tetratrichomonas* sp.	10.0 2.4 0.9 0.9 0.6 0.6	*Oochoristica* sp.	3.0	Oxyuridae Strongylida Ascarididae *Physaloptera* sp. Filarioidea *Capillaria* sp. *Strongyluris* sp. *Strongyloides* sp.	57.1 11.8 6.9 6.3 5.4 0.3 0.3 0.3	
	Chelonians	*Balantidium* sp. *Nyctotherus* sp.	26.2 1.6	Unidentified genus	0.3	*Tachygonetria* sp. Strongylida *Angusticaecum* sp. *Strongyloides* sp.	81.8 43.7 20.3 3.7	
Italy	Snakes	Eimeriidae	27	*Hymenolepis nana*	2	*Strongyloides* spp. *Capillaria* spp.	2 2	[6]
	Lizards	Eimeriidae	36	Unidentified	-	Oxyuridae	48	
Italy	Chelonians	*Nyctotherus* sp. *Cryptosporidium* spp. *Balantidium* sp. Coccidia Giardia spp.	18.6 12.6 0.9 0.9 0.5	Unidentified genus	0.5	Oxyuridae Ascarididae Strongylida	74.4 2.8 0.9	[10]
Germany	Chelonians	*Hexamita* sp. *Balantidium* spp. *Entamoeba* spp. *Trichomonas* spp. *Blastocystis* spp. *Hartmanella* spp. *Trimitus* spp.	0.007 0.007 0.005 0.004 0.002 0.001 0.001	Unidentified	-	Oxyuridae *Angusticaecum* spp. *Strongyloides* spp. *Heterakis* spp.	43.2 0.01 0.003 0.001	[5]

Table 1. *Cont.*

Countries	Reptiles	Protozoa	%	Cestodes	%	Nematodes	%	Reference
Germany	Lizards	*Isospora amphiboluri* *Entamoeba* spp. *Choleoeimeria* spp. *Trichomonas* spp. *Cryptosporidium* spp.	17 0.8 0.5 0.3 0.3	Unidentified	-	Oxyuridae *Strongyloides* spp.	41.2 0.1	[11]
Germany	All	Coccidia	4	Unidentified genus	1.6	Oxyuridae	18.7	[12]
		Cryptosporidium spp.	0.1			Ascarididae	5.2	
		Balantidium sp.	0.1			Strongylida	4.0	
						Hookworms	0.6	
						Strongyloides spp.	0.6	
						Capillaria sp	0.4	
						Lungworms	0.1	
Spain	Lizards	Coccidia	33	Unidentified	-	Oxyuridae	56	[13]
	Chelonians	*Nyctotherus* sp. *Balantidium* sp.	22 11	Unidentified	-	Oxyuridae	28	
Portugal	All	Coccidia *Isospora* spp. *Nyctotherus* spp. *Eimeria* spp.	23.0 10.8 5.0 2.9	Unidentified genus	0.7	Oxyuridae Strongylida Ascarididae *Strongyloides* spp. *Rhabdias* spp.	41.0 10.8 1.4 0.7	[14]
Portugal	Chelonians	*Nyctotherus* sp. *Amoeba* spp.	31.3 25.0	Unidentified	-	Oxyuridae	37.5	[15]
	Lizards	*Nyctotherus* sp.	25.0	Unidentified	-	Oxyuridae	75.0	
Portugal	Snakes	Eimeriidae	6.2	Unidentified	-	Nematoda	6.2	[16]
	Lizards	*Entamoeba* sp. *Balantidium* sp. Coccidia *Nyctotherus* sp. *Isospora amphiboluri* *Choleoeimeria* sp.	34.2 18.2 12.1 9.0 9.0 6.0	Unidentified	-	Nematoda	51.3	
	Chelonians	*Entamoeba* sp. *Balantidium* spp.	33.3 66.6	Unidentified	-	Pharyngodonidae	100	

There are several methods for quantitative parasitological analysis, such as McMaster, FLOTAC, and Mini-FLOTAC. The last method provides faster results with higher repeatability and sensitivity while requiring less labor work and no centrifugation [17,18]. Although techniques like Mini-FLOTAC can be applied to samples from any host, their use in reptiles is uncommon [3]. This study aims to assess the presence of gastrointestinal parasites in reptiles kept in a Portuguese zoo using two techniques: a flotation test (using a saturated sodium chloride—NaCl—solution) and Mini-FLOTAC.

2. Materials and Methods

Between February 2021 and February 2022, fecal samples were collected from reptiles in a Portuguese zoo. In zoos, the aim is to carry out minimal animal manipulation. To achieve this, most animals have never been dewormed or previously dewormed animals are not dewormed for at least a year. All the samples collected were from animals placed in individual enclosures. For this study, 99 reptile samples were collected from 22 species of reptiles kept at Zoo da Maia (Portugal) and analyzed for the presence of gastrointestinal parasites (Table 2).

Table 2. Species (scientific and common name) and number of reptiles examined.

Reptiles	Common Name	Number (N = 99)
	Chelonians/Order Testudines	
Testudo graeca	Greek tortoise	5
	Lizards/Suborder Lacertilia	
Pogona vitticeps	Bearded dragon	12
Phelsuma madagascariensis	Madagascar gecko	4
Uromastix geyri	Geyr's dabb lizard	4
Salvator merianae	Southern black-and-white teju	4
Ophisaurus apodus	Sheltopusik	4
Hemitheconyx caudicinctus	African gecko	2
Iguana iguana	Iguana	3
Zonosaurus maximus	Southeastern girdled lizard	3
Oplurus cuvieri	Collared iguana	1
	Snakes/Suborder Serpentes	
Lampropeltis getula californiae	California kingsnake	12
Python reticulatus	Reticulated python	7
Python bivittatus	Burmese python	6
Python regius	Ball python	6
Pantherophis obsoletus	Western rat snake	5
Epicrates angulifer	Cuban boa	5
Boa constrictor	Boa constrictor	4
Pantherophis guttatus	Corn snake	4
Lampropeltis triangulum hondurensis	Honduran milk snake	3
Lampropeltis getula splendida	Desert kingsnake	2
Morelia spilota	Carpet python	2
Epicrates cenchria maurus	Brown rainbow boa	1

The samples were collected from the environment immediately after defecation, during the normal daily activities of animal management and facility hygiene. After collection, the samples were stored in identified plastic bags at 4 °C and processed within 48 h through coprological methods in the laboratory of the Escola Superior Agrária of the Instituto Politécnico de Viana do Castelo (ESA—IPVC), Portugal. Each fecal sample was macroscopically checked for tapeworm proglottids and adult roundworms and then analyzed through a flotation test (a saturated sodium chloride solution, specific gravity of 1.2) [4] and Mini-FLOTAC techniques. Concerning Mini-FLOTAC, two flotation solutions were used—NaCl (specific gravity of 1.2) and zinc sulphate ($ZnSO_4$) (specific gravity of 1.35)—following the instructions reported in the original description by Cringoli et al. [19]. A dilution factor of 1:20 was used in Mini-FLOTAC (2 g of sample was added to 38 mL of solution). To the best of our knowledge, a Mini-FLOTAC protocol for reptiles has not yet been defined; therefore, only qualitative results were considered for both techniques: Mini-FLOTAC and the traditional flotation test. To detect the presence of eggs/oocysts, a conventional optical microscope was used with total magnifications of 100× and 400×. Eggs and oocysts were identified according to the descriptions of Zajac et al. [4].

3. Results

Of the 99 samples examined, we detected the presence of parasites (eggs/oocysts) in 53 (53.5%). All of the chelonians investigated presented parasites (100%, n = 5/5), while the prevalence of parasites decreased in lizards (56.8%, n = 21/37), followed by snakes (47.4%, n = 27/57) (Tables 3 and 4). Both techniques showed the same parasites for each sample.

Table 3. Species and number of lizards examined and frequency of parasites.

Species	*n* (N = 37)	Frequency of Parasites Number of Samples	%
Pogona vitticeps	12	11	91.7
Phelsuma madagascariensis	4	0	0.0
Hemitheconyx caudicinctus	2	2	100
Iguana iguana	3	2	66.7
Uromastix geyri	4	3	75.0
Salvator merianae	4	1	25.0
Zonosaurus maximus	3	0	0.0
Ophisaurus apodus	4	1	25.0
Oplurus cuvieri	1	1	100

Table 4. Species and number of snakes examined and frequency of parasites.

Species	*n* (N = 57)	Frequency of Parasites Number of Samples	%
Lampropeltis getula californiae	12	6	50.0
Python reticulatus	7	0	0.0
Python bivittatus	6	0	0.0
Python regius	6	2	33.3
Epicrates angulifer	5	4	80.0
Pantherophis obsoletus	5	5	100
Boa constrictor	4	4	100
Pantherophis guttatus	4	1	25.0
Lampropeltis triangulum hondurensis	3	2	66.7
Morelia spilota	2	1	50.0
Lampropeltis getula splendida	2	1	50.0
Epicrates cenchria maurus	1	1	100.0

Overall, the most frequent parasites were oxyurids (36.4%, $n = 36/99$) and strongylid/*Kalicephalus* sp. (8.1%, $n = 8/99$) (Table 5 and Figure 1).

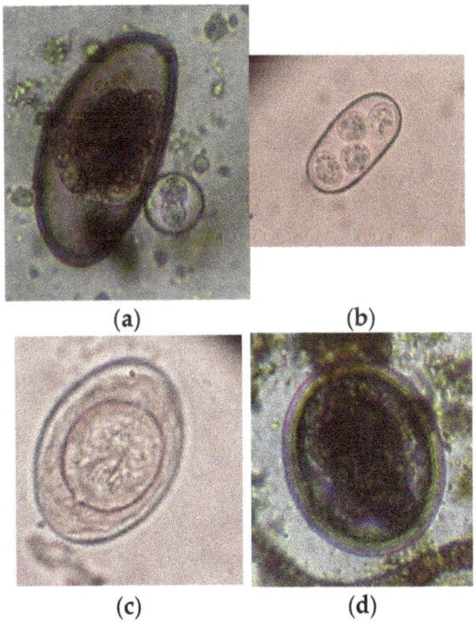

Figure 1. Microphotographs of reptile gastrointestinal parasites. (**a**) Oxyurid egg (on the left) and *Isospora* sp. oocyst (on the right); (**b**) *Eimeria* sp. oocyst; (**c**) *Hymenolepis nana* eggs; (**d**) Ascarid egg. All images were taken using a total magnification of 400×.

Table 5. Frequency of parasites in examined chelonians, lizards, and snakes.

Parasites	Chelonians (n = 5)	Lizards (n = 37)	Snakes (n = 57)	Frequency (n)	Frequency (%)
Oxyurids	1	21	14	36	36.4
Strongylids/*Kalicephalus* sp.	1	1	6	8	8.1
Eimeria sp.	0	0	5	5	5.1
Hymenolepis spp.	0	0	5	5	5.1
Ascarids	4	0	0	4	4.0
Isospora sp.	0	2	0	2	2.0

Among the samples where parasites were identified (n = 53), mixed parasitic infections were found in six (11.3%) reptiles. Only one chelonian out of the five investigated presented mixed parasitic infections (20.0%), followed by 2 lizards out of 37 (5.4%) and 3 snakes out of 57 (5.3%) investigated.

4. Discussion

Gastrointestinal parasites, primarily protozoa and nematodes, are commonly found in reptiles in captivity, possibly due to the monoxenous life cycle of several species and the high resistance of oocysts, eggs, or larvae, allowing them to survive in captive habitats [3]. Furthermore, as stated by Panayotova-Pencheva et al. [1], the confined environment of zoo habitats promotes geo-helminth development and maintenance, leading to frequent re-infections of reptiles.

According to Jacobson [20], parasites from the superfamily Oxyuroidea are usually observed as intestinal parasites in chelonians and lizards, often developing a commensal relationship with the host. The high frequency of oxyurids in this study in lizards and snakes can be explained by the fact that these parasites do not require an intermediate host to complete their life cycle [21]. Moreover, animals with an herbivorous and insectivorous diet are frequently parasitized by oxyurids [20], with at least twelve different genera of oxyurids already reported in snakes, lizards, and chelonians [21]. Moreover, the eradication of oxyurids may not always be successful [22]. Several studies reported a high prevalence of these parasites [2,5,6,10–15].

In the present study, ascarids were only identified in chelonians; however, they can also be found in snakes and lizards [2]. Ascarids have a monoxenous life cycle but can be transmitted via feeding or cohabitation through a paratenic host, such as amphibians, small mammals, and other reptiles [22]. Although the number of chelonian samples in this study was small (n = 5), the frequency of these parasites was high (80.0%).

The prevalence of strongylids (8.1%) in this study was lower compared to previous studies that found a prevalence of 20.4% (11/55), mainly *Kalicephalus* spp. [2], or 19.7% (14/74) [23]. Both of these previous studies were carried out in pet reptiles, which may explain the higher prevalence of parasites. The zoo sanitary and medical prophylaxis measures are more efficient in preventing parasitic infections. Conversely, other studies reported a lower prevalence, such as Pasmans et al. [3]. Although the definite host of *Kalicephalus* spp. is snakes [20], a small percentage of chelonians and lizards could also be affected (possibly functioning as paratenic hosts). These parasites have a direct life cycle and an oral infection route, with the possibility of transcutaneous infection, although this has not been properly proven [24].

Coccidiosis is one of the main causes of morbidity and mortality in reptiles [25]. The most common coccidian genera in these animals are *Eimeria* and *Isospora* [26], as found in this study. These parasites have a direct life cycle and are transmitted via the fecal–oral route through food, fomites, and infected substrates present in the terrarium [20,22], thus increasing the likelihood of transmission.

In this study, only two bearded dragons (*Pogona vitticeps*) were parasitized by the genus *Isospora*. Although the number is small, this is considered by Heard et al. [9] to be the most commonly reported coccidiosis in this species. Despite the small number, the prevalence of this genus in lizards (5.4%, 2/37) was higher than in the study by Rataj et al. [2], who reported 0.9%. These differences could be explained by the sample size of the different studies, as Rataj et al. [2] analyzed 949 samples while we investigated 99 samples.

According to Jacobson [20], the genus *Eimeria* is usually found in snakes and lizards. In this study, *Eimeria* was not identified in lizards, and in snakes, it had a lower prevalence (5.1%) compared to other studies (27.0% and 6.2%, respectively) [6,16]. The differences found between the studies may be related to the origin of the samples (whether they were collected from pet reptiles or a zoo), the size of the sample, and also the enormous diversity of reptiles that exist, each with its physiology, management, and very particular microbiota.

Although *Hymenolepis* sp. was found in this study in boas (*Boa constrictor*), its definitive host is the field vole [27]. Rinaldi et al. [6] considered it a pseudoparasite. In the present study, as boas are carnivorous, this parasite may have been acquired through feeding on live voles. Although it does not harm reptiles, this parasite is zoonotic [27], which underlines the importance of maintaining good hygiene practices among zoo staff during feeding and cleaning activities.

Overall, parasites can infect reptiles in a zoo through the introduction of a new animal that has not been quarantined in the zoo, through intermediary and paratenic hosts, or, even more probably, through food contamination and human visitors or staff who might be infected with these parasites [1]. All these parameters can and should be as controlled as possible, aiming for minimal manipulation of these animals (for instance, to proceed to medical prophylaxis or treatment) to avoid stress and for animal well-being purposes. This could be achieved by promoting training of zoo staff on biosecurity measures, zoonoses, and the importance of good personal hygiene practices to reduce the cross-contamination between habitats and protect animals and human health using a One Health perspective. Frequent and regular cleaning of the environment must be practiced, which will undoubtedly reduce the number of infective stages of different parasites in the environment, hence controlling such parasites.

The Mini-FLOTAC technique used in this study proved to be sensitive, as described in the literature. In the future, it would be useful to define a Mini-FLOTAC protocol for reptiles, together with reference values of eggs/oocysts per gram of feces that would allow distinguishing parasitism from parasitosis.

5. Conclusions

This study highlights the importance of assessing the presence of gastrointestinal parasites in reptile enclosures in zoological gardens. The present study found that 53.3% of the reptiles were infected with gastrointestinal parasites, demonstrating that, even with the high management standards practiced at the zoo, along with regular fecal examinations, a non-negligible frequency of parasitic infections remains. It is considered that if the biosecurity measures were reduced, a higher frequency of infection would become evident. Routine coprological examinations should always be conducted, and if diagnosed, animals should be properly treated and their environment sanitized as soon as possible. Furthermore, it is crucial to promote training of zookeepers from a One Health perspective.

Author Contributions: Conceptualization, D.D., C.L.B. and T.L.M.; methodology, D.D., C.L.B., M.J.C., N.A. and T.L.M.; formal analysis, D.D. and T.L.M.; investigation, C.L.B., M.J.C., N.A. and T.L.M.; resources, T.L.M.; data curation, D.D., M.J.C., C.L.B. and T.L.M.; writing—original draft preparation, D.D. and T.L.M.; writing—review and editing, D.D., C.L.B. and T.L.M.; supervision, T.L.M.; funding acquisition, T.L.M. All authors have read and agreed to the published version of the manuscript.

Funding: The participation of Teresa Letra Mateus was funded by national funds through FCT—Fundação para a Ciência e Tecnologia, I.P., within the scope of the project UIDB/05937/2020 with the DOI identifier https://doi.org/10.54499/UIDB/05937/2020 and UIDP/05937/2020 with the DOI identifier https://doi.org/10.54499/UIDP/05937/2020.

Institutional Review Board Statement: Ethical review and approval were waived for this study due to the fact that the samples were collected non-invasively; therefore, this study does not include any experimentation on animals.

Data Availability Statement: The original contributions presented in the study are included in the article; further inquiries can be directed to the corresponding author.

Acknowledgments: We would like to acknowledge the zookeepers who helped with the sampling of this study. The authors wish to thank Carla Canedo Ribeiro for her constructive feedback on earlier drafts. David Dantas, Cláudia Luís Batista, and Maria João Castro participated in this study in the context of their bachelor/master's degree in Veterinary Nursing at IPVC.

Conflicts of Interest: The authors declare no conflicts of interest.

References

1. Panayotova-Pencheva, M.S. Parasites in Captive Animals: A Review of Studies in Some European Zoos. *Zool. Gart.* **2013**, *82*, 60–71. [CrossRef]
2. Rataj, A.V.; Lindtner-Knific, R.; Vlahović, K.; Mavri, U.; Dovč, A. Parasites in pet reptiles. *Acta Vet. Scand.* **2011**, *53*, 33. [CrossRef] [PubMed]
3. Pasmans, F.; Blahak, S.; Martel, A.; Pantchev, N. Introducing reptiles into a captive collection: The role of the veterinarian. *Vet. J.* **2008**, *175*, 53–68. [CrossRef] [PubMed]
4. Zajac, A.M.; Conboy, G.A.; Little, S.E.; Reichard, M.V. *Veterinary Clinical Parasitology*, 9th ed.; Wiley-Blackwell: Hoboken, NJ, USA, 2021.
5. Hallinger, M.J.; Taubert, A.; Hermosilla, C.; Mutschmann, F. Occurrence of health-compromising protozoan and helminth infections in tortoises kept as pet animals in Germany. *Parasites Vectors* **2018**, *11*, 1–12. [CrossRef] [PubMed]
6. Rinaldi, L.; Mihalca, A.D.; Cirillo, R.; Maurelli, M.P.; Montesano, M.; Capasso, M.; Cringoli, G. FLOTAC can detect parasitic and pseudoparasitic elements in reptiles. *Exp. Parasitol.* **2012**, *130*, 282–284. [CrossRef] [PubMed]
7. Raiti, P. Snakes. In *BSAVA Manual of Exotic Pets*, 4th ed.; Meredith, A., Redrobe, S., Eds.; British Small Animal Veterinary Association: Quedgeley, UK, 2002; pp. 241–256.
8. Magnino, S.; Colin, P.; Dei-Cas, E.; Madsen, M.; McLauchlin, J.; Nöckler, K.; Prieto Maradona, M.; Tsigarida, E.; Vanopdenbosch, E.; Van Peteghem, C. Biological risks associated with consumption of reptile products. *Int. J. Food Microbiol.* **2009**, *134*, 163–175. [CrossRef] [PubMed]
9. Heard, D.; Fleming, G.; Lock, B.; Jacobson, E. Lizards. In *BSAVA Manual of Exotic Pets*, 4th ed.; Meredith, A., Redrobe, S., Eds.; British Small Animal Veterinary Association: Quedgeley, UK, 2002; pp. 223–240.
10. Nonnis, F.; Tamponi, C.; Pinna, S.; Diana, F.; Pudda, F.; Muzzeddu, M.; Cavallo, L.; Zeinoun, P.; Carta, C.; Varcasia, A.; et al. Epidemiological survey of gastrointestinal helminthes and protozoa in Testudines from Sardinia, Italy. *Vet. Parasitol. Reg. Stud. Andreports* **2024**, *54*, 101084. [CrossRef] [PubMed]
11. Hallinger, M.J.; Taubert, A.; Hermosilla, C.; Mutschmann, F. Captive Agamid lizards in Germany: Prevalence, pathogenicity and therapy of gastrointestinal protozoan and helminth infections. *Comp. Immunol. Microbiol. Infect. Dis.* **2019**, *63*, 74–80. [CrossRef] [PubMed]
12. Murnik, L.C.; Schmäschke, R.; Bernhard, A.; Thielebein, J.; Eulenberger, K.; Barownick, N.; Gawlowska, S.; Delling, C. Parasitological examination results of zoo animals in Germany between 2010 and 2022. *Int. J. Parasitol. Parasites Wildl.* **2024**, *24*, 100942. [CrossRef] [PubMed]
13. Ruivo, M.G.V.V. Parasitas Gastrointestinais em Répteis de Estimação em Barcelona. Master's Thesis, Universidade de Lisboa, Lisbon, Portugal, 2019. Available online: https://repositorio.ulisboa.pt/handle/10400.5/17612 (accessed on 3 February 2024).
14. Vítor, B.A.B. Prevalência de Parasitas Gastrointestinais em Répteis Domésticos na Região de Lisboa. Master's Thesis, Universidade Lusófona de Humanidades e Tecnologias, Lisbon, Portugal, 2018.
15. Bernardino, M.S.N. Parasitas Gastrointestinais de uma Colecção Privada de Geckos-Leopardo (*Eublepharis macularius*) e de Répteis Tidos como Animais de Estimação no Norte de Portugal. Master's Thesis, Universidade Lusófona de Humanidades e Tecnologias, Lisbon, Portugal, 2014. Available online: https://repositorio.ulisboa.pt/handle/10400.5/7616 (accessed on 3 February 2024).

16. Carvalho, T.J.F. Parasitas Gastrointestinais em Répteis de uma Coleção Comercial. Master's Thesis, Universidade de Lisboa, Lisbon, Portugal, 2018. Available online: https://repositorio.ulisboa.pt/handle/10400.5/16344 (accessed on 3 February 2024).
17. Szatmári, V. Feline Lungworm Infection. In *August's Consultations in Feline Internal Medicine*; Little, S., Ed.; Elsevier: St. Louis, MO, USA, 2015; Volume 7, pp. 379–386.
18. Barda, B.D.; Rinaldi, L.; Ianniello, D.; Zepherine, H.; Salvo, F.; Sadutshang, T.; Cringoli, G.; Clementi, M.; Albonico, M. Mini-FLOTAC, an Innovative Direct Diagnostic Technique for Intestinal Parasitic Infections: Experience from the Field. *PLoS Neglected Trop. Dis.* **2013**, *7*, e2344. [CrossRef] [PubMed]
19. Cringoli, G.; Maurelli, M.P.; Levecke, B.; Bosco, A.; Vercruysse, J.; Utzinger, J.; Rinaldi, L. The Mini-FLOTAC technique for the diagnosis of helminth and protozoan infections in humans and animals. *Nat. Protoc.* **2017**, *12*, 1723–1732. [CrossRef] [PubMed]
20. Jacobson, E. Parasites and Parasitic Diseases of Reptiles. In *Infectious Diseases and Pathology of Reptiles: Color Atlas and Text*; Jacobson, E., Ed.; Taylor & Francis Group: Boca Raton, FL, USA, 2007; pp. 571–666.
21. Taylor, M.; Coop, R.; Wall, R. Reptiles. In *Veterinary Parasitology*, 3rd ed.; Blackwell Publishing: Oxford, UK, 2007; pp. 1523–1543.
22. Greiner, E.C.; Mader, D.R. Parasitology. In *Reptile Medicine and Surgery*, 2nd ed.; Greiner, E.C., Mader, D.R., Eds.; Elsevier: Boston, MA, USA, 2006; pp. 343–364.
23. Papini, R.; Manetti, C.; Mancianti, F. Coprological survey in pet reptiles in Italy. *Vet. Rec.* **2011**, *169*, 207. [CrossRef] [PubMed]
24. Anderson, R. The Superfamily Diaphanocephaloidea. In *Nematode Parasites of Vertebrates, Their Development and Transmission*; Anderson, R., Ed.; CABI Publishing: Wallinford, UK, 2000; pp. 43–44.
25. Tully, J.T.; Mitchell, M. Reptiles and Amphibians. In *A Veterinary Technician's Guide to Exotic Animal Care*, 2nd ed.; Tully, J., Mitchell, M., Eds.; AAHA Press: Lakewood, OH, USA, 2012; pp. 37–78.
26. Greiner, E. Coccidiosis in reptiles. *Semin. Avian Exot. Pet Med.* **2003**, *12*, 49–56. [CrossRef]
27. Tijjani, M.; Majid, R.A.; Abdullahi, S.A.; Unyah, N.Z. Detection of rodent-borne parasitic pathogens of wild rats in Serdang, Selangor, Malaysia: A potential threat to human health. *Int. J. Parasitol. Parasites Wildl.* **2020**, *11*, 174–182. [CrossRef] [PubMed]

Journal of
Zoological and Botanical Gardens

MDPI

Article

Genetic Identification of Parasitic *Giardia enterica* in Three Wild Rodent Species from a Zoological Institution: First Host Records in Brazilian Porcupine (*Coendou prehensilis*) and Naked Mole Rat (*Heterocephalus glaber*), and Detection in Crested Porcupine (*Hystrix cristata*)

Lorena Esteban-Sánchez [1], Marta Mateo-Barrientos [1], Manuel de la Riva-Fraga [2], Lino Pérez de Quadros [2], Juan José García Rodríguez [1] and Francisco Ponce-Gordo [1,*]

[1] Department of Microbiology and Parasitology, Faculty of Pharmacy, Complutense University, Plaza Ramón y Cajal s/n, 28040 Madrid, Spain; lorees01@ucm.es (L.E.-S.); mmateo14@ucm.es (M.M.-B.); jjgarc01@ucm.es (J.J.G.R.)

[2] Veterinary Services, Parque Zoológico Faunia, 28032 Madrid, Spain; madelariva@faunia.es (M.d.l.R.-F.); lino@faunia.es (L.P.d.Q.)

* Correspondence: pponce@ucm.es

Abstract: Flagellates of the genus *Giardia* are intestinal parasites with a broad host range. Several *Giardia duodenalis* variants (assemblages) recently elevated to species rank—*G. duodenalis* (assemblage A1), *G. intestinalis* (A2) and *Giardia enterica* (B) are human pathogens. *Giardia enterica* has been reported in some hystricomorph rodents such as wild crested porcupines (*Hystrix cristata*), but no data were previously available from Brazilian porcupines (*Coendou prehensilis*) and naked mole rats (*Heterocephalus glaber*). The aim of this study is to genetically identify the *Giardia* isolates from these three rodent species, all housed in a zoological institution. Fecal samples were processed using the Bailenger concentration method, and DNA was extracted from the sediments using commercial kits. Partial PCR amplification and sequencing of the glutamate dehydrogenase, beta-giardin, and triose-phosphate isomerase genes revealed that all isolates belonged to *G. enterica*, showing 99–100% identity with sequences available in GenBank. Prevalences could not be reliably estimated due to small group sizes and the resulting proportions may be biased. To our knowledge, this is the first report identifying *Giardia* (*G. enterica*) in *C. prehensilis* and *H. glaber*, thus expanding the known host range of this parasite species and reinforcing the importance of surveillance in captive wild hosts.

Keywords: rodents; *Coendou prehensilis*; *Hystrix cristata*; *Hetercephalus glaber*; *Giardia enterica*; glutamate dehydrogenase; beta-giardin; triose-phosphate isomerase; PCR amplification and sequencing

1. Introduction

Rodents (order Rodentia) represent the most diverse mammal group, comprising nearly 2600 described species [1] accounting for more than 40% of currently living mammal species [2]. A defining feature of the group is a pair of open-rooted, chisel-shaped incisor teeth in each jaw, adapted for gnawing and requiring continuous wear to maintain functionality. The systematic of the order remains under revision; the suborders Sciuromorpha and Hystricomorpha are widely accepted, while a third suborder, Supramyomorpha, includes three infraorders (Myomorphi, Castorimorphi, and Anomaluromorphi) [1] which were previously classified as distinct suborders (Myomorpha, Castorimorpha and Anomaluromorpha) [2].

The Hystricomorpha, characterized by a hystricomorphous zygomasseteric system (in which the medial masseter muscle extends anteriorly through a greatly enlarged infraorbital foramen to insert on the lateral surface of the rostrum) [3] includes 17 families [1]. Among these are the Old World and New World porcupines and the African mole rats. A distinctive feature of porcupines is the presence of quills (modified hairs coated with thick keratin plates) located on the dorsal region and serving as a defense mechanism against predators. Old World porcupines (family Hystricidae), such as the crested porcupine (*Hystrix cristata*) and the Indian crested porcupine (*Hystrix indica*) are large rodents (reaching up to 27 kg in the crested porcupine). They are fossorial and of nocturnal habits. Primarily herbivorous, they can also consume insects and small vertebrates. They are frequently found in anthropogenic agricultural systems and are known to damage crops, thus being regarded as pests by farmers [4,5]. Consequently, they may play a role in the transmission of parasites to and from domestic animals and humans. In contrast, New World porcupines (family Erethizontidae) are smaller (the Brazilian porcupine, *Coendou prehensilis*, weighs approximately 4 kg), arboreal, and adapted to feeding in trees. In the wild, they have limited contact with humans; however, certain species, such as the Brazilian porcupine, are easily maintained in captivity in zoological gardens, increasing their potential contact with humans. With respect to the mole rats, this term refers to two families: Heterocephalidae (which includes a single genus and species, the naked mole rat *Heterocephalus graber*) and Bathyergidae. They are small, mouse-sized African rodents that inhabit complex underground tunnel systems and feed on tuberous plants. Their relevance to human interaction has increased in recent years due to their emergence as important animal models in biomedical research due to their eusocial behavior, unique ecophysiological adaptations, and exceptional longevity (up to 30 years in captivity) accompanied by sustained good health most of their lifespan [6,7].

Zoological institutions maintain a wide range of wild species under human supervision and care, facilitating closer contact between animals, zoo personnel, and visitors. These conditions can favor cross-species transmission of pathogens, particularly for parasites with direct life cycles, such as *Giardia* [8]. Monitoring gastrointestinal parasites in captive wildlife is therefore essential, not only for ensuring animal health but also for assessing zoonotic risks and preventing potential outbreaks [9]. This is especially relevant in zoological institutions, where the possibility of reverse zoonosis [9,10] may affect the health of captive populations and potentially impact wild populations if infected animals are released for reintroduction in the wild. Despite the importance of these issues, genetic characterization of *Giardia* isolates from captive zoo animals remains limited, particularly among lesser-studied rodent groups.

In routine parasitological monitoring of animals in zoological institutions in Madrid, Spain, *Giardia* cysts have been found in crested porcupines [11] and, more recently, in Brazilian porcupines and naked mole rats. The genus *Giardia* comprises diplomonadid flagellate protozoan parasites that infect a wide range of terrestrial vertebrates, including species parasitizing amphibians (*Giardia agilis*), reptiles (*Giardia varani*), birds (*Giardia ardeae, Giardia psittaci*) and mammals (*Giardia duodenalis sensu lato, Giardia muris, Giardia microti, Giardia peramelis*, and *Giardia cricetidarum*) [12]. These parasites have a direct life cycle, with eight-flagellated, binucleated trophozoites inhabiting the small intestine of the host, and four-nucleated cysts as the infective stage.

The taxonomy of *G. duodenalis* sensu lato has long been a subject of considerable debate, with three different names (*G. duodenalis*, *Giardia intestinalis*, and *Giardia lamblia*) being used in the scientific literature: *G. lamblia* has traditionally been favored in medical literature, whereas *G. duodenalis* has been more prevalent in non-medical contexts, with *G. intestinalis* generally regarded as a synonym in most cases [13]. The application of molecular tools has revealed extensive genetic variability within the species, leading to the identification of genetically distinct variants (assemblages) with varying host specificities, and ultimately to the proposal of several distinct species [14]: *G. duodenalis* sensu stricto (=*G. duodenalis* assemblage A1), *G. intestinalis* (=*G. duodenalis* A2), *Giardia enterica* (=*G. duodenalis* B), *Giardia canis* (=*G. duodenalis* C), *Giardia lupus* (=*G. duodenalis* D), *Giardia bovis* (=*G. duodenalis* E), *Giardia cati* (=*G. duodenalis* F), and *Giardia simoni* (=*G. duodenalis* G). Two additional assemblages (A3 and H) have been suggested as new species (*Giardia cervus* and *Giardia pinnipedis*, respectively), though they have not yet been formally proposed [14]. In citing identifications from previous studies, we will adopt this taxonomic framework and refer to these species names, rather than by assemblage letter designations. *Giardia duodenalis*, *G. intestinalis*, and *G. enterica* are all recognized as etiological agents of human giardiasis, a common gastrointestinal disease characterized by symptoms such as diarrhea, vomiting, and abdominal pain. Giardiosis is among the most prevalent gastrointestinal human infections worldwide, affecting an estimated 180 million people annually [15], and is responsible for numerous outbreaks in both low- and high-income countries [16].

In rodents, *G. muris*, *G. microti*, *G. simoni*, and *G. cricetidarum* have been identified through genetic analyses in species belonging to the Supramyomorpha-Myomorphi clade, primarily within murids and cricetids. However, these species have not been reported in members of Supramyomorpha-Castorimorphi, Hystricomorpha, or Sciuromorpha. In contrast, *G. duodenalis* sensu lato has been detected across all these clades, with a predominance in Hystricomorpha and Sciuromorpha (Table 1). For the family Hystricidae (Old World porcupines), genetic data are available only from a few wild individuals in Europe [17,18], while no molecular data are currently available for isolates from Erethizontidae (New World porcupines) or mole rats. The aim of the present study is to genetically identify *Giardia* isolates from these hosts. To our knowledge, this is the first time that *Giardia* has been genetically identified to the species level in the Brazilian porcupine and the African naked mole rats. These findings contribute to the epidemiological understanding of *Giardia* infections in rodents and help assess the potential role of these hosts as reservoirs for zoonotic transmission.

Table 1. Species of *Giardia* identified in rodent hosts by using molecular markers.

Rodent Order-Suborder	Family	Common Name	Rodent Genus	*Giardia* Species *	References
Sciuromorpha	Sciuridae	Squirrels	*Callosciurus*	*G. enterica*	[19]
			Spermophilus	*G. enterica*	[20]
		Chipmunks	*Eutamias*	*G. duodenalis*, *G. simoni*	[21]
		Marmots	*Marmota*	*G. duodenalis*, *G. enterica*, *G. bovis*	[20]
		Prairie dogs	*Cynomys*	*G. duodenalis*, *G. intestinalis*, *G. enterica*	[22]
Hystricomorpha	Caviidae	Guinea pigs Patagonian cavy	*Cavia* *Dolichotis*	*G. enterica*	[23]
	Chinchillidae	Chinchillas	*Chinchilla*	*G. duodenalis*, *G. intestinalis*, *G. enterica*, *G. canis*, *G. lupus*, *G. bovis*	[24–28]
	Echimyidae	Nutrias	*Myocastor*	*G. duodenalis*, *G. enterica*	[29]
		Hutias	*Capromys*	*G. enterica*	[26]
	Hystricidae	Old World porcupines	*Hystrix*	*G. enterica*	[17,18,30]

Table 1. *Cont.*

Rodent Order-Suborder	Family	Common Name	Rodent Genus	*Giardia* Species *	References
Supramyomorpha					
Myomorphi	Cricetidae	Hamsters	*Mesocricetus*	*G. cricetidarum*	[31]
		Dwart hamsters	*Phodopus*	*G. muris*, *G, cricetidarum*, *G. duodenalis* s.l.	[31]
		Voles	*Arvicola*	*G. microti*	[32]
			Clethrionomys	*G. microti*, *G. duodenalis*, *G. intestinalis*	[33]
			Eothenomys	*G. microti*	[31]
			Microtus	*G. microti*, *G. intestinalis*	[32–35]
			Myodes	*G. microti*, *G. duodenalis*/*G. intestinalis*, *G. enterica*	[33]
		Muskrats	*Ondrata*	*G. microti*, *G. enterica*, *G. canis*	[36,37]
		Deer mice	*Peromyscus*	*G. microti*	[38]
	Muridae	Mice	*Apodemus*	*G. microti*, *G. duodenalis* s.l.	[32,33,35]
			Mus	*G. microti*, *G. duodenalis* s.l., *G. simoni*	[23,32,39,40]
		Rats	*Niniventer*	*G. muris*	[35]
			Rattus	*G. microti*, *G. enterica*, *G. simoni*	[23,31,35,39,40]
	Spalacidae	Bamboo rats	*Rhizomys*	*G. enterica*	[41]
Castorimorphi	Castoridae	Beavers	*Castor*	*G. duodenalis*, *G. enterica*	[36,42]

* The identifications as *G. duodenalis* assemblages made by some authors have been renamed in the table as the corresponding species *G. duodenalis* (assemblage A1), *G. intestinalis* (A2), *G. enterica* (B), *G. canis* (C), *G. lupus* (D) or *G. simoni* (G). following Wielinga et al. (2023) [14]. When the assemblage is not indicated, the identification is noted as sensu lato (s.l.).

2. Materials and Methods

2.1. Sample Collection and Microscopical Analysis

The present study is based on *Giardia*-positive fecal samples collected in 2024 from crested porcupines, Brazilian porcupines, and naked mole rats housed at Faunia, a zoological park located in Madrid (Spain). These are the samples now analyzed to achieve the genetic identification of *Giardia* isolates in these host species. All animals were kept in isolated enclosures designed to mimic their natural habitats, with controlled lightning and humidity conditions. In all three species, animals were separated from visitors by transparent glass panels, preventing any direct visitor-animal contact.

Sample collection was performed by zookeepers early in the morning during routine cleaning and environmental maintenance procedures; no direct handling of the animals occurred during the collection of samples used in this study. Care was taken to avoid contamination with soil or plant debris. For the porcupine species, animal groups were formed by two individuals and fecal samples in each sampling event were obtained from both of them. In the case of the naked mole rats, which were housed as a colony of 11 individuals, pooled fecal boluses were collected from the "toilet chamber" of the colony. Samples were stored in clean, new plastic containers and transported to the laboratory within 1–3 h. after collection for immediate processing. Fecal concentrates were prepared using the sodium acetate-ether stool concentration technique [43]. Briefly, a small amount of each sample (2–3 g for the porcupine samples, 10–15 fecal pellets for the naked mole rats) was homogenized in 30 mL of acetate-acetic acid buffer (1.5% sodium acetate and 0.36% acetic acid in distilled water), then filtered through a metal sieve. The filtrate was mixed 1:1 with diethyl ether, vortexed for 30 s, and centrifuged at 1500 rpm for 2 min. The resulting aqueous sediment was examined microscopically on temporary slides, either unstained or stained with Lugol's iodine. Cysts were photographed and measured using an Olympus DP20 camera mounted on an Olympus BX51 microscope (Olympus, Tokyo, Japan).

2.2. DNA Extraction, Gene Selection and Amplification

One microscopically positive sample from each host species was further processed for genetic analysis. DNA was extracted using the Speedtools tissue DNA extraction kit (Biotools B&M Labs S.A., Madrid, Spain) following the manufacturer's instructions. Extracted DNA was either used immediately for PCR amplification or stored at −20 °C until used. Following Capewell et al. (2011) [44], partial fragments of four genes commonly used for *Giardia* genotyping were amplified by nested or semi-nested PCR and subsequently sequenced: small subunit rRNA (SSU-rRNA), beta-giardin (bg), triose phosphate isomerase (tpi) and glutamate dehydrogenase (gdh). All PCR amplifications were performed by using the PureTaq™ Ready-To-Go™ PCR beads (Merck KGaA, Darmstadt, Germany) in a final volume of 25 µL, containing 2 µL of each primer solution and 5 µL of template DNA (either total DNA extracted from the sample or the first-round PCR product). Reactions were carried out in an Eppendorf Master Cycler Gradient thermocycler (Eppedorf AG, Hamburg, Germany).

For the SSU-rRNA gene, the amplicon obtained from an initial PCR using forward primer Gia2029 (5′-AAGTGTGGTGCAGACGGACTC-3′) and reverse primer Gia2150c (5′-CTGCTGCCGTCCTTGGATGT-3′) was used as a template for a second (nested) PCR with forward primer RH11 (5′-CATCCGGTCGATCCTGCC-3′) and reverse primer RH4 (5′-AGTCGAACCCTGATTCTCCGCCAGG-3′), generating a 292 bp fragment [45]. The thermocycler was programmed as follows: initial denaturation at 96 °C for 4 min, followed by 35 cycles of 96 °C for 45 s denaturation), 55 °C (first PCR) or 59 °C (nested PCR) for 30 s (annealing), and 72 °C for 45 s (extension), with a final extension at 72 °C for 4 min [45].

Amplification of a 511 bp fragment of the bg gene was achieved using a primary PCR with forward primer G7 (5′-AAGCCCGACGACCTCACCCGCAGTGC-3′) and reverse primer G759 (5′-GAGGCCGCCCTGGATCTTCGAGACGAC-3′) [46], followed by a secondary PCR with forward primer 5′-GAACGAGATCGAGGTCCG-3′ and reverse primer 5′-CTCGACGAGCTTCGTGTT-3′ [47]. The thermocycling conditions were: initial denaturation at 95 °C for 15 min, 35 cycles of 95 °C for 30 s, 65 °C (primary PCR) or 55 °C (secondary PCR) for 30 s, and 72 °C for 1 min, with a final extension of 72 °C for 7 min [47].

For the tpi fragment (530 bp fragment), a primary amplification was performed using forward primer AL3543 (5′-AAATTATGCCTGCTCGTCG-3′) and reverse primer AL3546 (5′-CAAACCTTTTCCGCAAACC-3′), followed by a second amplification with forward primer AL3544 (5′-CCCTTCATCGGTGGTAACTT-3′) and reverse primer AL3545 (5′-GTGGCCACCACTCCCTGTCC-3′) [36]. PCR conditions for both rounds included an initial denaturation at 94 °C for 5 min, followed by 35 cycles of 94 °C for 45 s, 50 °C for 45 s, and 72 °C for 60 s, with a final extension of 72 °C for 10 min [36].

Amplification of a fragment of 432 bp of the gdh gene was conducted via an initial PCR with forward primer GDHeF (5′-TCAACGTYAAYCGYGGYTTCCGT-3′) and reverse primer GDHiR (5′-GTTRTCCTTGCACATCTCC-3′), followed by a semi-nested PCR using forward primer GDHiF (5′-CAGTACAACTCYGCTCTCGG-3′) and reverse primer GDHiR [48]. Thermocycling conditions for both PCR amplifications were: initial denaturation at 94 °C for 2 min, followed by 55 cycles of 94 °C for 30 s, 56 °C for 20 s, and 72 °C for 45 s, with a final extension of 72 °C for 7 min [48].

2.3. Sequence Analysis and Comparisons

In all cases, final PCR products obtained after the second (nested or semi-nested) amplification were visualized on 1% agarose gels stained with Pronasafe (Condalab, Torrejón de Ardoz, Spain) using a UV transilluminator (NuGenius Syngene, Cambridge, UK). Amplicons were purified using the QIAquick®® PCR Purification Kit (Qiagen, Hilden, Germany) and sequenced using an AbiPrism 3730XL DNA Analyzer (Applied Biosystems, now ThermoFisher Scientific, Waltham, MA, USA). Chromatograms were analyzed with ChromasPro ver. 2.1.10 (Technelysium Pty Ltd., South Brisbane, Australia) and compared against sequences in GenBank using the blastn algorithm available on the Nacional Center for Biotechnology Information website (https://blast.ncbi.nlm.nih.gov/Blast.cgi; last accessed on 21 May 2025).

Phylogenetic trees were constructed separately for each gene to enable comparative analyses between the sequences obtained in this study and those previously published from *Giardia* isolates in rodents. For this purpose, *Giardia* sequences from rodent hosts, as well as reference sequences for *G. duodenalis*, *G. intestinalis*, and *G. enterica* [14], including subtypes BIII and BIV of *G. enterica* [49], were retrieved from GenBank and aligned using the Muscle algorithm [50] implemented in MEGA-X [51]. For each gene, alignments were trimmed to match the length of the second-round amplicons; sequences covering less than 50% of alignment positions were excluded from the analysis. The best-fitting nucleotide substitution model was selected for each gene based on both the Akaike information criterion and the Bayesian information criterion, as calculated in MEGA-X. Phylogenetic trees were generated using the Neighbor-Joining method; boostrap resampling was performed 1000 times to assess branch support.

3. Results

3.1. Microscopical Findings

During 2024, several fecal samples tested positive for *Giardia* cysts: 3 out of 13 samples (23.1%) from crested porcupines, 3 out of 14 (21.4%) from Brazilian porcupines, and 1 out of 3 (33.3%) from naked mole rats. In both porcupine species, only one animal from each group (not always the same individual each time) was found to be infected at any given sampling event. For naked mole rats, the number of infected individuals could not be determined, as samples were collected as pooled feces and individual testing was not possible. In all cases, fecal consistency was normal and no clinical signs were observed in the animals.

The cysts exhibited typical morphological characteristics (Figure 1). They were observed in large numbers in the concentrated fecal sediments from crested porcupines, whereas they were scarce in the fecal concentrates from the Brazilian porcupine and the naked mole rat.

3.2. Genetic Analyses and Comparisons

PCR amplicons of the expected size were successfully obtained for all four genetic markers. However, the sequences corresponding to the SSU-rRNA fragments were illegible due to the presence of numerous double and triple peaks in the chromatograms. In contrast, the sequences obtained for the other markers were clear and unambiguous. Only in the case of the gdh marker was a consistent nucleotide ambiguity (Y = C/T) observed across all isolates, located at the same position as the ambiguity (R = G/A) in the GDHiR primer. The sequences obtained in this study have been deposited in the GenBank/EMBL/DDBJ databases under accession numbers PV391923-PV391925 (partial bg gene), PV391926-PV391928 (partial tpi gene) and PV391929-PV391931 (partial gdh gene).

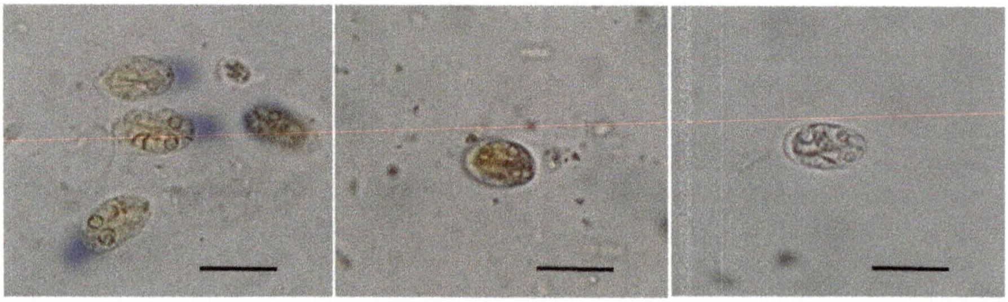

Figure 1. *Giardia* cysts observed on temporary slides stained with Lugol's iodine of the fecal concentrates from crested porcupines (**left**), Brazilian porcupine (**center**), and naked mole rat (**right**). Scale bar: 10 μm.

A comparison of the sequences obtained from the different host species revealed that they were not identical, with some nucleotide differences observed across all three markers (bg, tpi, and gdh) (Tables 2–4, Supplementary Files S1–S3). Comparisons with sequences available in GenBank confirmed that all isolates corresponded to *G. enterica* for each of the three markers. For the tpi gene, the sequences obtained were identical to previously deposited sequences. However, for the bg and gdh genes, novel sequence variants were identified (Table 5; Supplementary Files S1–S3). In phylogenetic analyses, the sequences clustered consistently within the *G. enterica* clade. A condensed tree based on the bg gene is shown in Figure 2 as representative of the results obtained for each gene; full phylogenetic trees for the bg, tpi, and gdh markers are provided in Supplementary Files S4–S6.

Table 2. Condensed alignment of beta-giardin sequences obtained from *Giardia* isolates of Brazilian porcupine, crested porcupine, and naked mole rat (highlighted in bold), compared with the reference *Giardia enterica* sequence (ACGJ01002392). Nucleotide positions are numbered according to the alignment shown in Supplementary File S1.

Sequence	Position		
	97	232	301
ACGJ01002392 *G. enterica* (strain GS/M)	R	T	C
AHHH01000111 *G. enterica* (strain GS)	G	T	C
PV391923 (Brazilian porcupine)	G	C	T
PV391924 (crested porcupine)	G	C	T
PV391925 (naked mole rat)	G	C	C

Table 3. Condensed alignment of triose-phosphate isomerase sequences obtained from *Giardia* isolates of Brazilian porcupine, crested porcupine, and naked mole rat (highlighted in bold), compared with the reference *Giardia enterica* sequences (ACGJ01002000 and AHHH01000009). Nucleotide positions are numbered according to the alignment shown in Supplementary File S2.

Sequence	Position										
	24	30	76	153	161	195	256	265	282	387	519
ACGJ01002000 *G. enterica* (strain GS/M)	A	T	T	T	G	A	C	A	A	A	T
AHHH01000009 *G. enterica* (strain GS)	A	T	T	T	G	A	C	A	A	A	T
PV391926 (Brazilian porcupine)	A	T	C	T	G	A	C	A	A	G	A
PV391927 (crested porcupine)	G	C	C	C	A	G	T	G	G	G	A
PV391928 (naked mole rat)	A	T	T	T	G	A	C	A	A	A	A

Table 4. Condensed alignment of glutamate dehydrogenase sequences obtained from *Giardia* isolates of Brazilian porcupine, crested porcupine, and naked mole rat (highlighted in bold), compared with the reference *Giardia enterica* sequences (ACGJ01002929 and AHHH01000018). Nucleotide positions are numbered according to the alignment shown in Supplementary File S3.

Sequence	Position				
	120	**309**	**333**	**405**	**429**
ACGJ01002929 *G. enterica* (strain GS/M)	C	C	C	C	T
AHHH01000018 *G. enterica* (strain GS)	C	C	C	C	T
PV391929 (Brazilian porcupine)	T	C	C	T	Y
PV391930 (crested porcupine)	C	A	C	C	Y
PV391931 (naked mole rat)	C	C	T	C	Y

Table 5. Comparisons of beta-giardin, triose phosphate isomerase, and glutamate dehydrogenase sequences obtained from *Giardia* isolates of Brazilian porcupine, crested porcupine, and naked mole rat (highlighted in bold) with reference sequences of *Giardia duodenalis*, *Giardia intestinalis* and *Giardia enterica* (highlighted in bold), and well as sequences available in GenBank using the blastn algorithm.

Beta-giardin Gene	*G. duodenalis* AACB03000002	*G. intestinalis* AHGT01000121	*G. enterica* ACGJ01002392	*G. enterica* AHHH01000111	Other Highly Similar *Giardia* Sequences
PV391923 (Brazilian porcupine)	93.94% (479/511)	93.43% (478/511)	99.61% (509/511)	99.61% (509/511) *	100.00% (511/511) AB618785 (human isolate) 100.00% (511/511) FJ009209 (anteater isolate)
PV391924 (crested porcupine)	93.74% (479/511)	93.43% (478/511)	99.61% (509/511)	99.61% (509/511) *	100.00% (511/511) AB618785 (human isolate) 100.00% (511/511) FJ009209 (from anteater)
PV391925 (naked mole rat)	93.93% (480/511)	93.73% (479/511)	99.80% (510/511)	99.80% (510/511) *	100.00% (511/511) KU504703 (human isolate) 100.00% (511/511) LC865371 (human isolate)
Triose-phosphate isomerase Gene	**AACB03000001**	**AHGT01000004**	**ACGJ01002000**	**AHHH01000009**	
PV391926 (Brazilian porcupine)	80.00% (424/530)	80.38% (426/530)	99.43% (527/530)	99.43% (527/530)	99.43% (527/530) LC865535 (human isolate) 99.43% (527/530) EU637591 (barbary macaque isolate)
PV391927 (crested porcupine)	80.57% (427/530)	80.94% (429/530)	97.92% (519/530)	97.92% (519/530)	99.25% (526/530) KM190834 (human isolate) 98.68% (523/530) MH310971 (human isolate)
PV391928 (naked mole rat)	79.62% (422/530)	80.94% (429/530)	99.81% (529/530)	99.81% (529/530)	99.81% (529/530) KM190822 (beaver isolate) 99.81% (529/530) HG970113 (human isolate)
Glutamate dehydrogenase Gene	**AACB03000002**	**AHGT01000014**	**ACGJ010022929**	**AHHH01000018**	
PV391929 (Brazilian porcupine)	90.28% (390/432) **	89.81% (388/432) **	99.31% (429/432) **	99.31% (429/432) **	99.77% (431/432) HM134209 (*Alouatta fusca* isolate) 99.77% (431/432) HM134211 (*Alouatta fusca* isolate)
PV391930 (crested porcupine)	90.28% (390/432) **	89.81% (388/432) **	99.31% (429/432) **	99.31% (429/432) **	99.54% (430/432) HM134215 (ostrich isolate) 99.54% (430/432) KM190702 (beaver isolate)
PV391931 (naked mole rat)	90.51% (391/432) **	90.05% (389/432) **	99.54% (430/432) **	99.54% (430/432) **	99.77% (430/431) LC430563 (human isolate) 99.54% (429/431) EF5076821 (human isolate)

* The ambiguity R (G/A) at position 97 of sequence AHHH01000111 is not considered a difference, as it is compatible with the nucleotides present at the corresponding position in the compared sequences. ** The ambiguity Y (C/T) at position 429 of sequences PV391929-PV391931 is not considered a difference, as it is compatible with the nucleotides present at the corresponding position in the reference sequences.

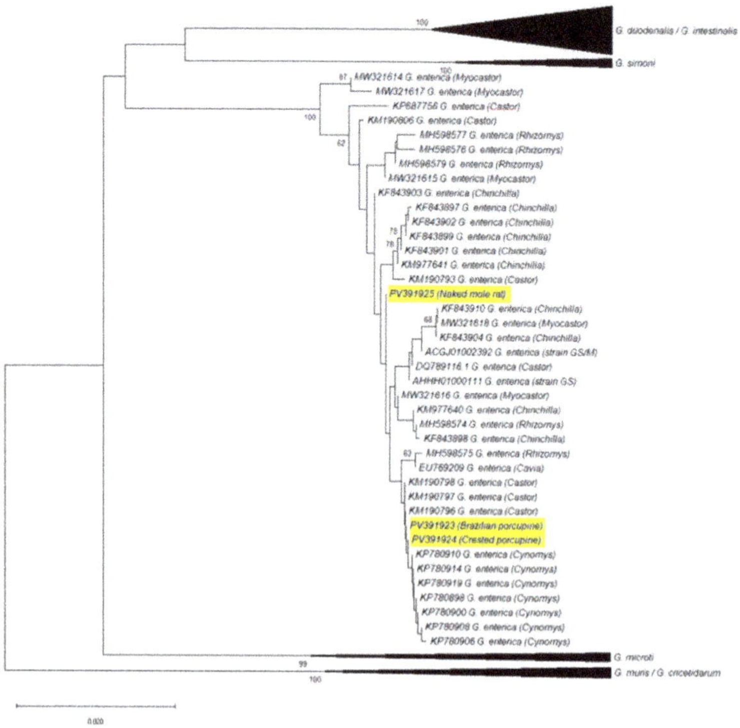

Figure 2. Condensed unrooted Neighbor-joining tree of the beta-giardin gene, constructed using the Tamura-Nei model of nucleotide substitution. Only the *Giardia enterica* clade is shown. Number at the nodes indicates bootstrap support values; only values greater than 60% are displayed. The sequences obtained in this study are indicated by a yellow-shaded box.

4. Discussion

The present results confirm that the crested porcupine can serve as a natural host for *G. enterica*, and provide the first genetic evidence of *G. enterica* infecting the Brazilian porcupine and the naked mole. Although this parasite is known to cause giardiasis in humans, no apparent clinical signs were observed in any of the infected rodents sampled in this study.

Zoo animals are particularly susceptible to infections by parasites with direct life cycles due to stress and the increased likelihood of encountering infective stages in confined environments. Additionally, there is potential for reverse zoonosis or "spill-back" transmission from humans (such as caretakers and visitors) to animals [9,10]. In a previous study [11], *Giardia* was detected in crested porcupines at a low frequency (8.5%) over a ten-year period, which increases to 13% when the current findings are included. The observed prevalence in the present study (50% in each positive sampling event) may be biased due to the small number of individuals sampled (n = 2), but it is comparable to prevalence in wild animals (38–100%) [30,52]. A similar limitation applies to the Brazilian porcupines, where only two individuals were sampled. Despite this, the prevalence observed (21.4% of samples testing positive over the year, and 50% of positive animals in the positive sampling events) is in line with previous findings in wild New World porcupines: in one study, one out of two *Coendou villosus* individuals tested positive for *Giardia* cysts [53], and in another, two out of ten individuals were infected [54]. In naked mole rats, determining prevalence would require individual handling of colony members, which was not conducted in the present study. The only previous report of *Giardia* in wild naked mole rats dates back several decades [55], describing a small number of trophozoites in the intestine and cysts in feces. To our knowledge, no other reports exist for *Giardia* infection in captive colonies of this species, nor in wild *Cryptomys* mole rats [56] or in captive *Fukomys damarensis* (Damaraland mole rats) [57].

The reasons for the failure to obtain readable sequences of the SSU-rRNA gene remain unclear. Nested or semi-nested PCR approaches are commonly employed to enhance the sensitivity and specificity of amplification, particularly when working with fecal DNA, which may contain low concentrations of target DNA, and PCR inhibitors [44]. In each case, the second-round PCR (nested or semi-nested) used a new set of internal primers and the product from the first round as a template, thereby minimizing the likelihood of non-specific amplification [36,45,47,48]. Given that high-quality sequences were obtained for the other markers (which are single-copy genes), it is unlikely that the SSU-rRNA gene amplification failure was due to low DNA concentration, poor DNA quality, or the presence of PCR inhibitors. It is therefore more plausible that the issue was related to amplicon purification or suboptimal sequencing conditions. Since the bg, gdh, and tpi markers yielded clear and informative sequences for genotyping, additional attempts to sequence the SSU-rRNA gene were not pursued, particularly in light of its lower discriminatory power for differentiating assemblages or species within the *G. duodenalis* complex [44]. The results obtained from the other gene markers are consistent with previous studies that identified *G. enterica* (*G. duodenalis* assemblage B) in Old World porcupines using SSU-rRNA, tpi, and bg gene sequences [17,18,30,58]. Notably, an infection by *G. intestinalis* (*G. duodenalis* assemblage A2) was also reported in wild crested porcupines in Italy [30], although the authors did not publish sequence data, preventing direct comparison.

Importantly, a comparative analysis of the sequences obtained from the three host species revealed that each was infected by a different *G. enterica* variant/genotype. In the case of the crested porcupine, the sequences differed from those previously reported in wild animals from Italy [17,18]. In natural environments, crested porcupines are exposed to a broad range of sympatric wildlife, including other reservoirs, and seasonal variation in habitat use and diet, all of which may influence both the risk of exposure and the selective pressures acting on circulating parasite strains. In contrast, zoo environments are characterized by increased host density, greater contact with humans (caretakers, visitors), artificial diets, reduced environmental microbial diversity, and potential cross-species transmission from other captive animals or contaminated sources. Such conditions may favor the establishment or persistence of different *Giardia* species and subspecific variants genotypes, or may facilitate the introduction and maintenance of variants not typically encountered in natural habitats. The present findings support this hypothesis and underscore the potential for captive environments to shape or concentrate specific parasite variants/genotypes of zoonotic relevance.

No genetic data are currently available for comparison in the case of the Brazilian porcupine and the naked mole rat. In Brazilian porcupines, previous reports have referred to the parasite as *Giardia* sp. [53,59] or as *G. intestinalis* [54], based solely on morphological identification. In naked mole rats, the parasite was identified as *Giardia cuniculi* [55], a junior synonym of *G. duodenalis* sensu lato. Both of these identifications in Brazilian porcupines [54] and naked mole rats [55] are compatible with the present identification as *G. enterica*.

Although the number of animals in the groups analyzed in this study is small, the objective was not to assess infection intensity or prevalence but rather to achieve molecular identification of *Giardia* isolates from zoo-housed rodent species that had not previously been genetically characterized. Accordingly, no attempts were made to quantify cysts in fecal samples, and as previously discussed, reported prevalence values may be biased. Notably, in both porcupine species, each group tested positive for *Giardia* on several occasions over time (although not always in the same individual), suggesting persistent infection within the group. Although only one positive sample per species was subjected to molecular analysis, and the possibility of additional genotypes circulating over time cannot be excluded, the sequence obtained from the crested porcupine matches those previously identified in wild individuals [17,18,30]. In contrast, the sequences from the isolates of Brazilian porcupine and naked mole rats represent the first available genetic data for *Giardia* in these host species. These findings provide a valuable reference for future comparative and epidemiological studies involving both captive and wild populations.

It is noteworthy that the sequences obtained from the three rodent species were not identical, indicating that different transmission events likely occurred; however, the specific routes of transmission remain unknown. The earliest detection of *Giardia* cysts in crested porcupine at this zoo was in 2020 [11], and in 2023 for the Brazilian porcupine [60]. All samples from naked mole rats had consistently tested negative until the current findings. It could be hypothesized that crested porcupines were initially infected and that the parasite was subsequently transmitted to other species through mechanical carriers (e.g., insects or fomites such as zookeepers' footwear), through infected local microfauna (e.g., small rodents), or potentially via infected zookeepers. However, given that distinct sequence variants were identified in each host species, it is more plausible that independent transmission events involving different *G. enterica* lineages occurred.

Transmission of the parasite from these animals to humans can be considered unlikely in the current zoo setting, as the animals are housed in isolated enclosures with minimal human contact. Only trained zookeepers enter the enclosures, and they do not directly interact with the animals except during veterinary check-ups. Enclosure cleaning is performed according to established protocols, and fecal material is disposed of under controlled conditions. Nevertheless, transmission risks cannot be completely excluded, and zoo animal-human transmission has been documented in other institutions [61]. In the present case, further investigation is needed within this zoological facility to clarify the potential routes of introduction and transmission of the parasite. Additionally, studies in porcupines and naked mole rats from other zoological institutions and wild populations are needed to better assess their potential role as zoonotic reservoirs of *Giardia* spp.

While further research and future discoveries may reveal additional *Giardia* genotypes or species in previously unreported hosts, there appears to be an association between *G. muris*, *G. microti*, and *G. cricetidarum* and rodents of the suborder Myomorphi. In contrast, within other rodent taxa, the predominant (and in some cases, the only) species identified belong to the *G. duodenalis* sensu lato complex, which is also capable of infecting myomorphic rodents. To date, no *Giardia* species have been reported in rodents of the suborder Anomaluromorphi (springhares, scaly-tailed flying squirrels, and *Zenkerella*).

5. Conclusions

This study provides the first genetic identification of *G. enterica* in the Brazilian porcupine and the naked mole rat, and confirms the presence of this parasite in captive individuals of the crested porcupine. Although all isolates were identified as *G. enterica* through three independent genetic markers (tpi, gdh, and bg), the sequences differed among host species and from sequences available in GenBank, suggesting independent infection events and previously undescribed variants. These findings support the hypothesis that captive environments may influence the circulation and selection of specific *Giardia* genotypes, which may differ from those found in wild populations.

No clinical signs of giardiasis were observed in infected individuals, and in both porcupine species, repeated positive results over time suggest persistent colonization. Given the zoonotic potential of *G. enterica*, ongoing surveillance in both captive and wild rodent populations is warranted, particularly to clarify transmission dynamics and evaluate the role of these hosts as reservoirs. The genetic data generated in this study provide a valuable reference for future epidemiological, comparative, and taxonomic research on *Giardia* in porcupines and mole rats.

Supplementary Materials: The following supporting information can be downloaded at https://www.mdpi.com/article/10.3390/jzbg6020028/s1, File S1: Alignment of beta-giardin sequences of *Giardia* species from rodent isolates. File S2: Alignment of triose-phosphate isomerase sequences of *Giardia* species from rodent isolates. File S3: Alignment of glutamate dehydrogenase sequences of *Giardia* species from rodent isolates. File S4: Neighbor-joining tree based on the partial beta-giardin gene sequences of *Giardia* species infecting rodents. File S5: Neighbor-joining tree based on the partial triose-phosphate isomerase gene sequences of *Giardia* species infecting rodents. File S6: Neighbor-joining tree based on the partial glutamate dehydrogenase gene sequences of *Giardia* species infecting rodents.

Author Contributions: Conceptualization, L.E-S., M.M.-B. and F.P.-G.; methodology, L.E.-S., M.M.-B. and F.P.-G.; formal analysis, L.E.-S. and F.P.-G.; investigation, L.E.-S. and F.P.-G.; resources, M.d.l.R.-F., L.P.d.Q., J.J.G.R. and F.P.-G. data curation, L.E.-S. and F.P.-G.; writing—original draft preparation, L.E.-S., M.M.-B. and F.P.-G.; writing—review and editing, L.E.-S., M.M.-B., M.d.l.R.-F., L.P.d.Q., J.J.G.R. and F.P.-G.; supervision, F.P.-G.; project administration, J.J.G.R. and F.P.-G.; funding acquisition, J.J.G.R. and F.P.-G. All authors have read and agreed to the published version of the manuscript.

Funding: This research was funded by the Research Group no. 911120 "Epidemiology, diagnostic and antiparasitic therapy" of the Complutense University.

Institutional Review Board Statement: Ethical approval was not required for this study. Fecal samples were collected non-invasively from the ground in the animals' enclosures, as part of routine monitoring procedures conducted by the zoo veterinary staff. No animal was captured, handled, or subjected to any intervention for the purposes of this research. All procedures complied with the park's institutional guidelines for animal welfare, ensuring minimal disturbance to the animals throughout the process.

Data Availability Statement: The original contributions presented in this study are included in the article, in the main text, or as Supplementary Materials. DNA sequences have been deposited in the GenBank/EMBL/DDBJ repositories under accession numbers PV391923-PV391931. Further inquiries about data supporting the conclusions of this article can be directed at the corresponding author.

Acknowledgments: We wish to thank all the staff of Faunia Park for their help with the collection, handling, and transportation of the samples. We wish to thank Manuela Pumar Martín for her help with processing the samples.

Conflicts of Interest: The authors declare no conflicts of interest.

Abbreviations

The following abbreviations are used in this manuscript:

SSU-rDNA	Small subunit ribosomal DNA gene
bg	Beta-giardin gene
tpi	Triose-phosphate isomerase gene
gdh	Glutamate dehydrogenase gene

References

1. D'Elía, G.; Fabre, P.-H.; Lessa, E.P. Rodent systematics in an age of discovery: Recent advances and prospects. *J. Mammal.* **2019**, *100*, 852–871. [CrossRef]
2. Carleton, M.D.; Musser, G.G. Order Rodentia. In *Mammal Species of the World. A Taxonomic and Geographic Reference*; Wilson, D.E., Reeder, D.M., Eds.; Johns Hopkins University Press: Baltimore, MD, USA, 2005; pp. 745–752. ISBN 978-0-8018-8221-0. Available online: https://www.departments.bucknell.edu/biology/resources/msw3/ (accessed on 21 March 2025).
3. Hautier, L.; Michaux, J.; Marivaux, L.; Vianey-Liaud, M. Evolution of the zygomasseteric construction in Rodentia, as revealed by a geometric morphometric analysis of the mandible of *Graphiurus* (Rodentia, Gliridae). *Zool. J. Linn. Soc.* **2008**, *154*, 807–821. [CrossRef]
4. Hafeez, S.; Khan, G.M.; Ashfaq, M.; Khan, Z.H. Food habits of the Indian crested porcupine (*Hystrix indica*) in Faisalabad, Pakistan. *Pak. J. Agric. Res.* **2011**, *48*, 205–210.
5. Franchini, M.; Frangini, L.; Viviano, A.; Filacorda, S.; Stokel, G.; Mori, E. Get out from my field! The role of agricultural crops in shaping the habitat selection by and suitability for the crested porcupine in Central Italy. *Mammal Res.* **2024**, *69*, 411–421. [CrossRef]
6. Buffenstein, R.; Park, T.; Hanes, M.; Artwohl, J.E. Naked mole rat. In *The Laboratory Rabbit, Guinea Pig, Hamster, and Other Rodents*; Suckow, M.A., Stevens, K.A., Wilson, R.P., Eds.; Elsevier Inc.: London, UK, 2012; pp. 1055–1074. [CrossRef]
7. Patterson, B.D.; Upham, N.S. A newly recognized family from the Horn of Africa, the Heterocephalidae (Rodentia: Ctenohystrica). *Zool. J. Linn. Soc.* **2014**, *172*, 942–963. [CrossRef]
8. Applebee, A.J.; Thompson, R.C.A.; Olson, M.E. *Giardia* and *Cryptosporidium* in mammalian wildlife—Current status and future needs. *Trends Parasitol.* **2005**, *21*, 370–376. [CrossRef]
9. Ryan, U.; Cacciò, S.M. Zoonotic potential of *Giardia*. *Int. J. Parasitol.* **2013**, *43*, 943–956. [CrossRef] [PubMed]
10. Al Noman, Z.; Tasnim, S.; Masud, R.I.; Anika, T.T.; Islam, M.S.; Rahman, A.M.M.T.; Rahman, M.T. A systematic review on reverse-zoonosis: Global impact and changes in transmission patterns. *J. Adv. Vet. Anim. Res.* **2024**, *11*, 601–607. [CrossRef]
11. Esteban-Sánchez, L.; García-Rodríguez, J.J.; García-García, J.; Martínez-Nevado, E.; de la Riva-Fraga, M.A.; Ponce-Gordo, F. Wild Animals in Captivity: An Analysis of Parasite Biodiversity and Transmission among Animals at Two Zoological Institutions with Different Typologies. *Animals* **2024**, *14*, 813. [CrossRef]
12. Ryan, U.M.; Feng, Y.; Fayer, R.; Xiao, L. Taxonomy and molecular epidemiology of *Cryptosporidium* and *Giardia*—A 50 year perspective (1971–2021). *Int. J. Parasitol.* **2021**, *51*, 1099–1119. [CrossRef]
13. Thompson, R.C.A.; Hopkins, R.M.; Homan, W.L. Nomenclature and genetic groupings of *Giardia* infecting mammals. *Parasitol. Today* **2000**, *16*, 210–213. [CrossRef] [PubMed]
14. Wielinga, C.; Williams, A.; Monis, P.; Thompson, R.C.A. Proposed taxonomic revision of *Giardia duodenalis*. *Infect. Genet. Evol.* **2023**, *111*, 106430. [CrossRef]
15. Torgerson, P.R.; Devleesschauwer, B.; Praet, N.; Speybroeck, N.; Willingham, A.L.; Kasuga, F.; Rokni, M.B.; Zhou, X.-N.; Fèvre, E.M.; Sripa, B.; et al. World Health Organization estimates of the global and regional disease burden of 11 foodborne parasitic diseases, 2010: A data síntesis. *PloS Med.* **2015**, *12*, e1001920. [CrossRef] [PubMed]
16. Krumrie, S.; Capewell, P.; Smith-Palmer, A.; Mellor, D.; Weir, W.; Alexander, C.L. A scoping review of risk factors and transmission routes associated with human giardiasis outbreaks in high-income settings. *Curr. Res. Parasitol. Vector-Borne Dis.* **2022**, *2*, 1000084. [CrossRef] [PubMed]
17. Procesi, I.G.; Di Filippo, M.M.; De Liberato, C.; Lombardo, A.; Brocherel, G.; Perrucci, S.; Di Cave, D.; Berrilli, F. *Giardia duodenalis* in wildlife: Exploring genotype diversity in Italy and across Europe. *Pathogens* **2022**, *11*, 105. [CrossRef]
18. Pacini, M.I.; Mazzei, M.; Sgorbini, M.; D'Alfonso, R.; Papini, R.A. A One-Year Retrospective Analysis of Viral and Parasitological Agents in Wildlife Animals Admitted to a First Aid Hospital. *Animals* **2023**, *13*, 931. [CrossRef]
19. Beck, R.; Sprong, H.; Bata, I.; Lucinger, S.; Pozio, E.; Cacciò, S.M. Prevalence and molecular typing of Giardia spp. in captive mammals at the zoo of Zagreb, Croatia. *Vet. Parasitol.* **2011**, *175*, 40–46. [CrossRef]
20. Xu, J.; Liu, H.; Jiang, Y.; Jing, H.; Cao, J.; Yin, J.; Li, T.; Sun, Y.; Shen, Y.; Wang, X. Genotyping and subyyping of *Cryptosporidium* spp. and *Giardia duodenalis* isolated from two wild rodent species in Gansu Province, China. *Sci. Rep.* **2022**, *12*, 12178. [CrossRef]
21. Deng, L.; Luo, R.; Liu, H.; Zhou, Z.; Li, L.; Chai, Y.; Yang, L.; Wang, W.; Fu, H.; Zhong, Z.; et al. First identification and multilocus genotyping of *Giardia duodenalis* in pet chipmunks (*Eutamias asiaticus*) in Sichuan Province, southwestern China. *Parasit. Vectors* **2018**, *11*, 199. [CrossRef]
22. Roellig, D.M.; Salzer, J.S.; Carroll, D.S.; Ritter, J.M.; Drew, C.; Gallardo-Romero, N.; Keckler, M.S.; Langham, G.; Hutson, C.L.; Karem, K.L. Identification of *Giardia duodenalis* and *Enterocytozoon bieneusi* in an epizoological investigation of a laboratory colony of prairie dogs, *Cynbomys ludovicianus*. *Vet. Parasitol.* **2015**, *210*, 91–97. [CrossRef]
23. Lebbad, M.; Mattsson, J.G.; Christensson, B.; Ljungström, B.; Backhans, A.; Andersson, J.O.; Svärd, S. From mouse to moose: Multilocus genotyping of *Giardia* isolates from various animal species. *Vet. Parasitol.* **2010**, *168*, 231–239. [CrossRef] [PubMed]

24. Levecke, B.; Meulemans, L.; Dalemans, T.; Casaert, S.; Claerebout, E.; Geurden, T. Mixed *Giardia duodenalis* assemblage A, B, C and E infections in pet chinchillas (*Chinchilla lanigera*) in Flanders (Belgium). *Vet. Parasitol.* **2011**, *177*, 166–170. [CrossRef]
25. Veronesi, F.; Fioretti, D.P.; Morganti, G.; Bietta, A.; Moretta, I.; Moretti, A.; Traversa, D. Occurrence of *Giardia duodenalis* infection in chinchillas (*Chinchilla lanigera*) from Italian breeding facilities. *Res. Vet. Sci.* **2012**, *93*, 807–810. [CrossRef] [PubMed]
26. Pantchev, N.; Broglia, A.; Paoletti, B.; Globokar Vrhovec, M.; Bertram, A.; Nöckler, K.; Cacciò, S.M. Occurrence and molecular typing of *Giardia* isolates in pet rabbits, chinchillas, guinea pigs and ferrets collected in Europe during 2006–2012. *Vet. Rec.* **2014**, *175*, 18. [CrossRef]
27. Qi, M.; Yu, F.; Li, S.; Wang, H.; Luo, N.; Huang, J.; Zhang, L. Multilocus genotyping of potentially zoonotic *Giardia duodenalis* in pet chinchillas (*Chinchilla lanigera*) in China. *Vet. Parasitol.* **2015**, *208*, 113–117. [CrossRef]
28. Gherman, C.M.; Kalmár, Z.; Györke, A.; Mircean, V. Occurrence of *Giardia duodenalis* assemblages in farmed long-tailed chinchillas *Chinchilla lanigera* (Rodentia) from Romania. *Parasit. Vectors* **2018**, *11*, 86. [CrossRef] [PubMed]
29. Cui, Z.; Wang, D.; Wang, W.; Zhang, Y.; Jing, B.; Xu, C.; Chen, Y.; Qi, M.; Zhang, L. Occurrence and multi-locus analysis of *Giardia duodenalis* in coypus (*Myocastor coypus*) in China. *Pathogens* **2021**, *10*, 179. [CrossRef]
30. Coppola, F.; Maestrini, M.; Berrilli, F.; Procesi, I.G.; Felicioli, A.; Perrucci, S. First report of *Giardia duodenalis* infection in the crested porcupine (*Hystrix cristata* L.; 1758). *Int. J. Parasitol: Parasites Wildl.* **2020**, *11*, 108–113. [CrossRef]
31. Lyu, Z.; Shao, J.; Xue, M.; Ye, Q.; Chen, B.; Qin, Y.; Wen, J. A new species of *Giardia* Künstler, 1882 (Sarcomastigophora: Hexamitidae) in hamsters. *Parasit. Vectors* **2018**, *11*, 202. [CrossRef]
32. Lux, L.; Ulrich, R.G.; Santos-Silva, S.; Querós, J.; Imholt, C.; Klotz, C.; Paupério, J.; Pita, R.; Vale-Gonçalves, H.; Alves, P.C.; et al. Detection and molecular characterization of *Giardia* and *Cryptosporidium* spp. circulating in wild small animals from Portugal. *Animals* **2023**, *13*, 515. [CrossRef]
33. Helmy, Y.A.; Spierling, N.G.; Schmidt, S.; Rosenfeld, U.M.; Reil, D.; Imholt, C.; Jacob, J.; Ulrich, R.G.; Aebischer, T.; Klotz, C. Occurrence and distribution of *Giardia* species in wild rodents in Germany. *Parasit. Vectors* **2018**, *11*, 213. [CrossRef] [PubMed]
34. De Liberato, C.; Moltabano Di Filippo, M.; Sagrafoli, D.; Ferraro, D.; Guadano Procesi, I.; Berrillli, F. *Giardia microti* in pete *Microtus guentheri*: Evidence of a parasite never detected in Italy. *Parasitol. Int.* **2021**, *80*, 102207. [CrossRef] [PubMed]
35. Feng, K.; Yang, S.; Xu, Y.; Wen, L.; Chen, J.; Zhang, W.; Chen, S.; Shen, Y.; Xiao, L.; Guo, Y.; et al. Molecular characterization of *Cryptosporidium* spp., *Giardia* spp. and *Enterocytozoon bieneusi* in eleven wild rodent species in China: Common distribution, extensive genetic diversity and high zoonotic potential. *One Health* **2024**, *18*, 100750. [CrossRef] [PubMed]
36. Sulaiman, I.M.; Fayer, R.; Bern, C.; Gilman, R.H.; Trout, J.M.; Schantz, P.M.; Das, P.; Lal, A.A.; Xiao, L. Triosephosphate isomerase gene characterization and potential zoonotic transmission of *Giardia duodenalis*. *Emerg. Infect. Dis.* **2003**, *9*, 1444–1452. [CrossRef]
37. Adriana, G.; Zsuzsa, K.; Oana, D.M.; Mircea, G.C.; Viorica, M. *Giardia duodenalis* genotypes in domestic and wild animals from Romania identified by PCR-RFLP targeting the gdh gene. *Vet. Parasitol.* **2016**, *217*, 71–75. [CrossRef]
38. Kilonzo, C.; Li, X.; Vodoz, T.; Xiao, C.; Chase, J.A.; Jay-Russell, M.T.; Vivas, E.J.; Atwill, E.R. Quantitative shedding of multiple genotypes of *Cryptosporidium* and *Giardia* by deer mice (*Peromyscus maniculatus*) in a major agricultural region of the California Central Coast. *J. Food Prot.* **2017**, *80*, 819–828. [CrossRef]
39. Zhao, Z.; Wang, R.; Zhao, W.; Qi, M.; Zhao, J.; Zhang, L.; Li, J.; Liu, A. Genotyping and subtyping of *Giardia* and *Cryptosporidium* isolates from commensal rodents in China. *Parasitology* **2015**, *142*, 800–806. [CrossRef]
40. Asghari, A.; Motazedian, M.H.; Asgari, Q.; Shamsi, L.; Sarkari, B.; Shahabi, S.; Mohammadi-Ghalehbin, B. Occurrence, genetic characterization, and zoonotic importance of *Giardia duodenalis* in various species of rodents (*Mus musculus*, *Rattus norvegicus*, and *Rattus rattus*). *Comp. Immunol. Microbiol. Infec. Dis.* **2022**, *85*, 101812. [CrossRef]
41. Ma, X.; Wang, Y.; Zhang, H.-J.; Wu, H.-X.; Zhao, G.-H. First report of *Giardia duodenalis* infection in bamboo rats. *Parasit. Vectors* **2018**, *11*, 520. [CrossRef]
42. Fayer, R.; Santín, M.; Trout, J.M.; DeStefano, S.; Koenen, K.; Kaur, T. Prevalence of Microsporidia, *Cryptosporidium* spp., and *Giardia* spp. in beavers (*Castor canadensis*) in Massachusetts. *J. Zoo Wildl. Med.* **2006**, *37*, 492–497. [CrossRef]
43. Bailenger, J. Mechanisms of parasitical concentration in coprology and their practical consequences. *J. Am. Med. Technol.* **1979**, *41*, 65–71.
44. Capewell, P.; Krumrie, S.; Katzer, F.; Alexander, C.L.; Weir, W. Molecular epidemiology of *Giardia* infections in the genomic era. *Trends Parasitol.* **2021**, *37*, 142–153. [CrossRef] [PubMed]
45. Appelbee, A.J.; Frederick, L.M.; Heitman, T.L.; Olson, M.E. Prevalence and genotyping of *Giardia duodenalis* from beef calves in Alberta, Canada. *Vet. Parasitol.* **2003**, *112*, 289–294. [CrossRef]
46. Cacciò, S.M.; De Giacomo, M.; Pozio, E. Sequence analysis of the β-giardin gene and development of a PCR-RFLP assay to genotype *Giardia duodenalis* cysts from human fecal samples. *Int. J. Parasitol.* **2002**, *32*, 1023–1030. [CrossRef]
47. Lalle, M.; Pozio, E.; Capelli, G.; Bruschi, F.; Crotti, D.; Cacciò, S.M. Genetic heterogeneity at the β-giardin locus among human and animal isolates of *Giardia duodenalis* and identification of potentially zoonotic subgenotypes. *Int. J. Parasitol.* **2005**, *35*, 207–213. [CrossRef] [PubMed]

48. Read, C.M.; Monis, P.T.; Thompson, R.C.A. Discrimination of all genotypes *of Giardia duodenalis* at the glutamate dehydrogenase locus using PCR-RFLP. *Infect. Genet. Evol.* **2004**, *4*, 125–130. [CrossRef]
49. Monis, P.T.; Andrews, R.H.; Mayrhofer, G.; Ey, P.L. Molecular systematics of the parasitic protozoa *Giardia intestinalis*. *Mol. Biol. Evol.* **1999**, *16*, 1135–1144. [CrossRef]
50. Edgar, R.C. MUSCLE: Multiple sequence alignment with high accuracy and high throughput. *Nucleic Acids Res.* **2004**, *32*, 1792–1797. [CrossRef]
51. Kumar, S.; Stecher, G.; Li, M.; Knyaz, C.; Tamura, K. MEGA X: Molecular Evolutionary Genetics Analysis across Computing Platforms. *Mol. Biol. Evol.* **2018**, *35*, 1547–1549. [CrossRef]
52. Viviano, A.; Huffman, M.A.; Senini, C.; Mori, E. Do porcupines self-medicate? The seasonal consumption of plants with antiparasitic properties coincides with that of parasite infections in *Hystrix cristata* of Central Italy. *Eur. J. Wildl. Res.* **2022**, *68*, 72. [CrossRef]
53. Soares, J.F.; Schafer da Silva, A.; Belmonte Oliveira, C.; Kipper da Silva, M.; Mariscano, G.; Salomão, E.L.; Gonzalez Monteiro, S. Parasitismo por *Giardia* sp. e *Cryptosporidium* sp. em *Coendou villosus*. *Ciência Rural. Santa Maria* **2008**, *38*, 548–550. [CrossRef]
54. Lallo, M.A.; Pereira, A.; Araujo, R.; Favorito, S.E.; Bertolla, P.; Bondan, E.F. Ocurrência de *Giardia, Cryptosporidium* e microsporidios em animais silvestres em área de desmatamento no Estado de São Paulo, Brasil. *Cienc. Rural.* **2009**, *39*, 1465–1470. [CrossRef]
55. Porter, A. Morphology and affinities of Entozoa and Endophyta of the naked mole rat *Heterocephalus glaber*. *Proc. Zool. Soc. London* **1957**, *128*, 515–527. [CrossRef]
56. de Graaf, G. On the parasites associated with the Bathyergidae. *Koedoe* **1964**, *7*, a806. [CrossRef]
57. Buffenstein, R.; Smith, M.; Amoroso, V.G.; Patel, T.V.; Ross, M.; Bassanpal, S.; Park, T.J.; Delaney, M.A.; Adams, C.R.; Arroyo, J.; et al. A new laboratory research model: The Damaraland mole-rat and its managed care. *J. Am. Assoc. Lab. Anim. Sci.* **2024**, *63*, 683–693. [CrossRef]
58. Garcia-R, J.C.; French, N.; Pita, A.; Velathanthiri, N.; Shrestha, R.; Hayman, D. Local and global genetic diversity of protozoan parasites: Spatial distribution of *Cryptosporidium* and *Giardia* genotypes. *PLoS Negl. Trop. Dis.* **2017**, *11*, e0005736. [CrossRef]
59. Moré, G.; Rüegg-van den Broek, P.; Glardon, O.J.; Gliga, D.S.; Frey, C.F.; Basso, W. Identification and management of infections caused by *Giardia* sp., *Trichuris* sp. and *Demodex* sp. in captive Brazilian porcupines. *Int. J. Parasitol. Parasites Wildlf.* **2024**, *25*, 100976. [CrossRef]
60. Esteban-Sánchez, L. (Complutense University, Madrid, Spain); de la Riva-Fraga, M. (Veterinary Services, Faunia Zoological Park, Madrid, Spain); Pérez de Quadros, L. (Veterinary Services, Faunia Zoological Park, Madrid, Spain); García-Rodríguez, J.J. (Complutense University, Madrid, Spain); Ponce-Gordo, F. (Complutense University, Madrid, Spain). *Giardia* spp. infections in the animals housed at the Faunia Zoological park (Madrid, Spain) during 2023–2024. Unpublished work.
61. Köster, P.; Martínez-Nevad, E.; González, A.; Abelló-Poveda, M.T.; Fernández-Bellon, H.; de la Riva-Fraga, M.; Marquet, B.; Guéry, J.-P.; Kanuf-Witzens, T.; Weigold, A.; et al. Intestinal protists in captive non-human primates and their handlers in six European zoological gardens. Molecular evidence of zoonotic transmission. *Front. Vet. Sci.* **2022**, *8*, 819887. [CrossRef]

MDPI AG
Grosspeteranlage 5
4052 Basel
Switzerland
Tel.: +41 61 683 77 34

Journal of Zoological and Botanical Gardens Editorial Office
E-mail: jzbg@mdpi.com
www.mdpi.com/journal/jzbg

www.ingramcontent.com/pod-product-compliance
Lightning Source LLC
Chambersburg PA
CBHW062055220926
51850CB00010B/2773
* 9 7 8 3 7 2 5 8 4 7 9 5 2 *